Communicating in Geography and the Environmental Sciences

Dr Iain Hay

Series editors
Deirdre Dragovich
Alaric Maude

Melbourne
OXFORD UNIVERSITY PRESS
Oxford Auckland New York

OXFORD UNIVERSITY PRESS AUSTRALIA

Oxford New York
Athens Auckland Bangkok Bogotá
Buenos Aires Calcutta Cape Town Chennai
Dar es Salaam Delhi Florence Hong Kong
Istanbul Karachi Kuala Lumpur Madrid
Melbourne Mexico City Mumbai Nairobi
Paris Port Moresby São Paulo Singapore
Taipei Tokyo Toronto Warsaw

and associated companies in
Berlin Ibadan

First published 1996
Reprinted 1997, 1998, 1999

National Library of Australia
Cataloguing-in-Publication data:

Communicating in geography and the environmental sciences

Bibliography.
Includes index.
ISBN 0 19 553942 7.

1. Learning and scholarship—Authorship. 2. Academic writing. 3. Study skills. 4. Geography—Study and teaching (Higher)—Australia. 5. Environmental sciences—Study and teaching (Higher)—Australia. I. Hay, Iain. II. Title. (Series: Meridian, Australian geographical perspectives).

378.1702812

Edited by Jenny Missen
Text designed by Perdita Nance
Typeset by Perdita Nance
Printed through Bookpac Production Services, Singapore
Published by Oxford University Press,
253 Normanby Road, South Melbourne, Australia

Foreword

Australian geographers have produced some excellent books in recent years, several of them in association with the 1988 bicentennial of European settlement in the continent, and all of them building on the maturing of geographical research in this country. However, there is a continuing need for relatively short, low-cost books written for university students, books that fill the gap between chapter-length surveys and full-length books and that explore the geographical issues and problems of Australia and its region, or present an Australian perspective on global geographical processes.

Meridian: Australian Geographical Perspectives is a series initiated by the Institute of Australian Geographers to fill this need. The term 'meridian' refers to a line of longitude linking points in a half-circle between the poles. In this series it symbolises the interconnections between places in the global environment and global economy, one of the key themes of contemporary geography

Titles in the series cover a range of topics representing contemporary Australian teaching and research in geography—for example, economic restructuring, vegetation change, land degradation, cities, natural hazards, and urban biophysical environments. Future topics include gender and geography, cultural geography, coastal management, and environmental impact assessment. Although the emphasis in the series is on Australia, forthcoming publications in the series will include occasional titles on South-East Asia, drawing on the considerable expertise developed by Australian geographers in relation to this region.

This book is part of an initiative to produce titles dealing with some of the skills that are an integral part of a geographical education. *Communicating in Geography* is the eighth title in the meridian series, and the first on skills. The book introduces students to the wide range of ways in which information and ideas are communicated in geography and the environmental sciences; from essays to posters, and from talks to examinations. By explaining the purpose of each method, and the likely expectations of the audience and the assessors, the author demystifies the criteria used to assess good communication, helping students to develop the ability to communicate clearly and effectively. Equipped with a better understanding of what constitutes good communication, students can go on to create their own styles of presentation.

Iain Hay has developed a strong interest and considerable expertise in geographical education. He has a Graduate Certificate in Tertiary Education, is the Australasian Commissioning Editor for the *Journal of Geography in Higher Education*, and has received two National Teaching Development Grants for research into geographical education. In 1995 he won a Flinders University Vice-Chancellor's Award for Excellence in Teaching. He has a reputation for innovation in his own teaching, and for expecting high standards from his students. As an example of good communication, this book meets those high standards, and we expect that it will be welcomed by both students and their teachers.

Deirdre Dragovich
University of Sydney

Alaric Maude
Flinders University

Contents

Figures

Tables

Acknowledgements

During this book's decade-long gestation period, a large number of people have contributed to work that was eventually to become part of the pages which follow. I am indebted to them all and am particularly grateful for the contributions, support and model teaching of Jane Abbiss, Alice Bass, Ken Bardsley, Jim Bell, Cecile Cutler, Ed Delaney, Debbie Faulkner, David Hodge, Erick Howenstine, Alaric Maude, and the hundreds of students at the University of Washington, Western Washington University, University of Wollongong and Flinders University who put up with my teaching and assessment experiments. I would also like to thank the Australian Committee for the Advancement of University Teaching for providing funds which supported parts of this project. Finally, to my parents—gifted teachers and learners—I owe special thanks.

Introductory comments

This book is about communicating effectively in academic settings. It discusses the character and practice of some of the most common forms of academic presentation skills used by students of geography and the environment. Chapters outline the 'whys' and 'hows' of essays, research and laboratory reports, reviews, summaries, annotated bibliographies, maps, figures, tables, posters, examinations, and talks. Information on the ways in which these forms of presentation are commonly assessed is another important part of the book.

Knowledge, information, and ideas remain the most highly valued currencies in universities. However, without the ability and means to communicate clearly and effectively, the value of one's thoughts and abstractions can be severely eroded. For that reason effective communication is a vital component of intellectual endeavour. One important ingredient of effective communication is an appreciation of the ways in which audiences make sense of the messages conveyed. Typically, audiences expect that certain conventions will be upheld or followed by people communicating to them through specific media. For instance, readers of an academic paper will usually expect some early introductory advice of the paper's purpose. People reviewing a scientific research report anticipate that information on supporting literature, research methods, and results will be set out in a customary order and will offer specific sorts of information. Unfortunately, however, many students do not know the accepted cues, clues, ceremonies, conventions, and characteristics associated with formal (academic) communication. In other words, some students do not know how to 'make the grade'. It is primarily because of that problem and because of the lack of specific, relevant advice to address it that this book was written.

In the pages which follow I have tried to demystify the conventions of communication associated with university exercises. By laying bare the criteria markers typically seek when evaluating specific forms of communication like essays, posters and talks, the book lets everyone know how grades are made. Just as importantly, the following pages set out the means by which grade-making criteria may be fulfilled.

The book serves a number of other important purposes. It is intended to:

- *Help improve teaching, learning and assessment within educational contexts of scarce resources.* Many academics now find themselves being asked to do more with less—to do more teaching, more research, more administration, and more community service with less money, less time, less public recognition, and less government support than ever before. One means of coping with this set of tensions is to teach more efficiently and effectively. This book is an attempt to contribute to those ends.
- *Help increasingly diverse student populations fulfil educational objectives.* Recent moves to mass tertiary education in many western countries have brought to universities students with far more diverse educational and cultural backgrounds than ever before. By revealing the characteristics of academic communication and assessment this book represents one attempt to accommodate that shift. In serving this purpose the book may also be of value to those offering and undertaking distance education courses.
- *Provide students with useful vocational skills.* In the 1990s, and in the context of emerging patterns of work and work organisation, the importance of university graduates' capacities to communicate ideas and information surfaces repeatedly in the reports of government and business think-tanks and academic authors. In some recognition of vocational considerations, this book makes an effort to contribute to the development of communication skills; however, its focus is traditional academic communication skills and principles, not business-related communication skills such as preparing a curriculum vitae or writing effective memoranda, press releases, and facsimile messages.
- *Codify and transfer teaching experience.* The book is one distillation of the experience and expertise possessed by those many people I have troubled over the years for comments on marking practice and communication conventions. With this book in hand, many new teachers and part-time teachers may find their teaching and assessment experiences simplified.

Most of the chapters have been written around a common framework comprising four parts. First, there is an explanation of the specific type of communication being discussed. This takes the form of an answer to a question such as 'Why prepare a poster?' This section is followed by a broad, conceptual statement of the key matters assessors seek when marking the particular type of communication under consideration. For instance, lecturers marking book reviews typically seek clearly expressed description, analysis, and evaluation of the text. An essay marker wishes to be told clearly what the author thinks and has learned about a specific topic. An outline of means of achieving effective contact with one's audience constitutes the third part of most chapters or sections. Where possible, this discussion is structured around an explicit statement of the sorts of criteria assessors have typically been found to use when marking student work. A statement of those assessment criteria forms the fourth and final part of most chapters. Lecturers may find these lists useful as marking guides (e.g. to ensure that a broad range of matters is considered during assessment; to offer consistency in marking practice) and they may be of considerable benefit to students seeking either a checklist which might be used in critical self-review or an indication of the sorts of things assessors are considering when marking work.

... for student readers

If you are like many students in most universities, you have a mountain of textbooks to read, thousands of photocopied words to digest, and dog-eared collections of handwritten notes to absorb. On top of all that material directly related to your course, you now have a book on academic communication skills. Do not be discouraged. It is true that there is a lot to read and absorb in the pages that follow, but it is unlikely that you will have to use all of the material in this book in any single semester university course. Instead, this is a book written to be used throughout your entire degree (and for some post-degree experiences too). For example, in a first year course, you might find the chapters on essay writing, graphics and exams most helpful. In third year, you might use the material on oral presentations for the first time while still using the material on essays and graphics. So, think of the book as comprising sections that will be important over three or four years, not just for one semester.

You might want to regard the book as something of a statement of achievement too. Let me explain that rather cryptic comment. If you use this text in the next few years at university, you should be able to complete a degree with a sound appreciation of almost all of the material in the book. Look at the book as an indicator of proficiency 'this is what I will know' and not as something which says, 'this is what I have to know'.

The following pages discuss in detail the conventions of communicating effectively in an undergraduate academic setting. In part, the advice given is based on reviews of patterns of marking and comment by academic staff. The book was written to provide you with an insight to some of the expectations of people for whom you will be writing essays, giving talks, and drawing figures. Those expectations are reviewed fully in the chapters but they are also summarised in the assessment schedules. An understanding of your audience's expectations **before** you undertake an assignment ought to help you prepare better work than might otherwise have been possible.

You will find it helpful to review the assessment schedules (and the appropriate chapters) before you begin an assignment requiring some specific form of communication (e.g. writing an essay, giving a talk). You will then be able to undertake the assignment with an understanding of the appropriate conventions of communication. When you have finished a draft of your work, try marking it yourself using the checklist as a guide. If you have a patient and thoughtful friend, ask them if they will do this for you too. The checklist will help to ensure that you and your friend give consideration to the broad range of issues likely to be examined by any assessor, whether or not they actually use the schedules in their own marking practice. If something in the assessment schedule does not make sense to you, consult the material in the appropriate chapter for an explanation. In this way you may be able to illuminate and correct any shortcomings in your work before those problems are uncovered by your lecturer. The end results of this process ought to be better communication and better grades.

One thing needs to be stressed from the outset. The following pages are intended to offer advice only and not prescriptions for 'perfect' assignments. The guidelines are intended to assist you in preparing assignments, types of which you might be undertaking for the first time (e.g. an academic poster, a formal talk). As well as heeding the content of this book, you should read journal articles and other people's essays critically, pay attention to the ways in which effective and poor speakers present themselves and their material, and be critical of maps and graphs. See what works and what does not. Learn from your observations and attempt to forge your own, distinctive approach of communicating. The style you develop may be very effective and yet transgress some of the guidelines set out in the following pages. This should not be a matter of concern: your individuality and imagination are to be celebrated and encouraged, not condemned and excised.

... for lecturers

This is not a book of magic. I do not wish to make wild claims about the benefits that might flow from student and teacher use of the pages that follow. However, I do have reason to believe that if material from this book is referred to and incorporated into teaching **and** assessment practice within a discipline or across a degree, there is likely to be an improvement in student communication skills. In my experience, a number of simple and effective strategies involving use of this book have yielded useful results.

- Before the class undertakes an exercise make available to students copies of the assessment schedules associated with most chapters.
- Encourage students, by whatever means you consider appropriate, to critically read assessment sheets and explanatory notes in the appropriate chapter of the book before they begin an assigned task.
- Apply strategies that encourage students to use marking schedules to assess their own work before submitting it for peer review or final assessment. Not only does this offer an opportunity for students to engage in critical self-reflection, but if submitted with work for instructor assessment, self-assessment sheets can be used as a 'diagnostic' tool, highlighting differences between students' perceptions of their own work and those of the assessor(s).
- Use peer review methods such as writing groups and student assessment of oral presentations as means of encouraging students to think critically about communication and the positions and expectations of author and audience. The assessment sheets associated with most of the chapters in this book are useful frameworks for peer review.
- Use the assessment sheets, and use them repeatedly, as the foundation for your own assessment practice.

These strategies, singly and in combination, while requiring little extra teaching effort, offer the potential for improvements in written, oral and graphic communication skills. Why not give them a try?

1

Writing essays

True ease in writing comes from art, not chance,
As those move easiest who have learned to dance.

Alexander Pope

This chapter briefly argues the case for writing before going on to discuss how to write a good essay. Most of the material in the chapter is devoted to a review of those matters your essay markers might be looking for when they are assessing your work. Much of the information and advice in the following pages is structured around the essay assessment schedule included at the end of the chapter.

Why Write?

You might think that essays and other forms of written work demanded by your lecturer are some sort of miserable torture inflicted on you as a part of an ancient academic initiation ritual. To tell the truth, however, there are some very good reasons to develop expertise in writing.

- It is an *academic and professional responsibility* to write. As an academic or practising geographer (perhaps labelled economist, planner, demographer) or environmental manager you should make the results of your work known to the public, to government, to sponsoring agencies, and the like. There is not only a moral obligation to make public the results of scientific inquiry, but you will probably be required to write—and to write well—as part of any occupation you take up.
- Writing is one of the most *powerful means we have of communicating*. It is also the most common means by which formal transmission of ideas and arguments is achieved.
- Among the most important reasons for writing is the fact that writing is a *generative, thought-provoking process*.

> *I write because I don't know what I think until I read what I have to say.*
> - FLANNERY O'CONNOR
>
> *You write—and find you have something to say.*
> - WRIGHT MORRIS
>
> *But I really write to find out about something and what is known about something ...I write books to find out about things.*
> - DAME REBECCA WEST

As these quotes suggest, writing promotes original thought. It also reveals how much you have understood about a particular topic.

- Writing is also a means by which you can *initiate feedback on your own ideas*. Through the circulation of your writing in forms such as professional reports, essays, letters to the editor, and journal or magazine articles you may spark replies that contribute to your own knowledge as well as to that of others. Writing (and other forms of communication) is critical to the development and re-shaping of knowledge.
- By forcing you to marshal your thoughts and present them coherently to other people, writing is also a *central part of the learning process.*
- Writing is a means of *conveying and creating the ideas of new worlds.* Writing is part of the process by which we give meaning to the world(s) in which we live. One of the ways in which we make sense of our world(s) is through the communications of others (e.g. journalists). In consequence, those who have power over communication have power over thought and, hence, power over reality. Control of communication is control over destiny. Without the ability to communicate effectively, you may be abandoning your own capacity for self-determination.
- *Writing can be fun.* Think of writing as an art form or as storytelling. Use your imagination. Paint the world you want with words.

How Do I Write a Good Essay?

Understanding and applying the conventions of effective writing

Writing a good essay requires that you anticipate and fulfil the demands and expectations of readers. The following pages outline some of those expectations. Fulfilling those demands also requires that you devote sufficient time to writing.

Devote sufficient time to research and writing

There is no formula for calculating the amount of time which needs to be devoted to writing essays of any particular length or 'mark value'. Some writers do their best work under great pressure of time; others work more slowly and may require several weeks to write a short essay. However, irrespective of your writing style, doing the research for a good essay does take time. So, do yourself a favour and devote plenty of time to finding and reading books and journal articles and to consulting other sources germane to your essay.

Practise writing

As with almost any art—or sport for that matter—practice improves your ability to perform. Practice allows you to apply the conventions of effective writing. It also provides you with the opportunity to seek feedback on the quality of your writing.

Seek and apply feedback

Writing is a *social* process. Perhaps you have an image of a good writer sitting alone at a keyboard typing an error-free, comprehensible, and publishable first draft of a manuscript. Sadly, that image is unfounded. Almost every writer produces countless drafts and seeks comment from peers and other reviewers. Even the late great F. Scott Fitzgerald had manuscripts sent back from editors for revision—after he had completed extensive revisions of his own.

When you write, you are writing for an audience. It is vital that you understand the ways in which those people might react to your work. A valuable means of gaining such understanding is through allowing friends, tutors, and others to read and comment on draft copies of your work. You might even find it useful to form a group with some friends and agree to proof-read one another's essays critically.

What are Your Essay Markers Looking For?

> Your lecturers are not looking for 'correct answers'. There is no 'line' for you to follow. They are concerned with how well you make your case. Whether they agree or disagree with your judgment is not essential to your mark. Disagreement does not lead to bad marks; bad essays do. (Lovell & Moore 1992, p. 4)

The answer to the question 'what are your markers looking for?' is really quite simple. Assessors want

> to be told clearly what you *think*
> and what you have *learned*
> about a *specific topic.*

The following guidelines and advice, which are written to match the criteria outlined in the assessment schedule at the end of this chapter, ought to help you satisfy the broad objective noted in the shaded box above. It is worth thinking seriously about these guidelines. During your degree program, you are likely to write about 40–50 essays, totalling about 100 000 words—twice as many words as there are in this book. You might as well spend a little time now ensuring that the 1000–2000 hours you spend writing those essays are as productive and rewarding as possible.

Quality of argument

> The argument fully addresses the question

If any one issue in particular can be identified as critical to a good essay, it is this one. Failure to address the question assigned or chosen is often a straightforward indicator of a lack of understanding of course material. It may also be seen as an indicator of carelessness in reading the question or of a lack of interest and diligence.

Completely answering the question requires a careful interrogation of the essay topic (e.g. what does 'describe' mean? How about 'analyse' or 'compare and contrast'? What do other key words in the assigned topic actually mean? (See the Glossary for a discussion of terms used commonly in essay assignments.) In defining the topic and the terms within it, discussion with other people in your class may be revealing. See how your friends have interpreted the question. In what ways do your views differ or overlap? Why? Take care here. While it is useful to seek help from your friends, do not devalue your own opinions. Your friends may be wrong! Stand up for your own viewpoint, but listen critically to the arguments of others and be prepared to adjust your opinion if you believe the evidence warrants it.

Wherever necessary, you should clarify the meaning of an assigned topic with your lecturer. Try to do this after you have given the topic full thought, discussed it with friends, and established your own interpretation, but before you begin writing your paper.

When you have completed your essay, use the paragraph listing and rearranging technique outlined in the next section to check that the organisation of your work reflects full coverage of the required material. Ask yourself if you have really dealt with the main points you wished to make.

Logically developed argument

> Nothing is more frustrating than to be lost in someone else's intellectual muddle. A paper that fails to define its purpose, that drifts from one topic to the next, that 'does not seem to go anywhere,' is certain to frustrate the reader. If that reader happens to be your lecturer, he or she is likely to strike back with notations scrawled in the margin criticising the paper as 'poorly organised,' 'incoherent,' 'lacking clear focus,' 'discursive,' 'muddled,' or the like. Most lecturers have developed a formidable arsenal of terms that express their frustration at having to wade through papers that...are poorly conceived or disorganised. (Friedman & Steinberg 1989, p. 53)

In assessing an essay, markers will usually look for a coherent framework of thought underpinning your work. They are trying to uncover the conceptual skeleton upon which you have hung your ideas and to see if it is orderly and logical. Throughout the essay readers need to be reminded of the connections between your discussion and the framework. Make clear the relationship between the point being made and the argument you are advancing.

For some topics—but not all—an essay framework can be formed before writing begins (see Essay plan and Sketch diagram discussions below). In other situations, the essay may take shape as it is being written (see Freewriting below).

Essay plan—if you can, try to set out an essay plan before you begin writing. That is, work out a series of broad headings that will form the framework upon which your essay will be constructed. Then, add increasingly detailed material under those headings until your essay is written. As you proceed, you may find it necessary to make changes to the overall structure of the essay.

Sketch diagrams of the subject matter are also a good means of working out the structure of your essay. Write down key words associated with the material you will discuss and draw out a sketch of the ways in which those points are connected to one another. Rearrange the diagram until you have formulated an outline. This can then be used in much the same way as an essay plan.

Freewriting—if you encounter 'writer's block', or are writing on a topic that does not lend itself to use of an essay plan, brainstorm and without hesitation write anything related to the topic until you have some paragraphs on the screen or page in front of you. Then remove the rubbish and organise the material into some coherent package. To be effective, this writing style requires a good background knowledge of the material to be discussed. Freewriting can be a good way of making connections between elements of the material you have read about. It is not an easy option for people who have not got a clue about their essay topic.

When the first draft of the essay is finally written, check the structure. This can be done quite easily. Go through the document giving each paragraph or section a heading that summarises that section's content. Then write out the headings on a separate sheet of paper or on cards. Read through the headings. Are they in a logical order? Do they address the assigned topic in a coherent fashion? If not, rearrange the headings until they do make sense and add new headings that might be necessary to cover the topic adequately. If additional headings are required, you will also have to write some new sections of your paper. Of course, you may also find that some sections can be removed. Make your amendments and then go through this process of assigning and arranging headings again until you are satisfied that the essay has a logical argument. Then, of course, rearrange the written material according to the new sequence of headings.

You might find that the summary of headings you have prepared supplies a framework upon which an informative introduction can be based. A reader provided with a sense of direction early in the paper should find your work easy to follow.

Writing well-structured through introduction, body, and conclusion

In almost all cases, good academic writing will have an introduction, a discussion, and a conclusion. In many cases too, one can conceive of an essay as taking the form of an hourglass. The introduction provides a broad outline, setting the topic in its context. The central discussion tapers in to cover the detail of the specific issue(s) you are exploring. The conclusion sets your findings back into the context from which the subject is derived and may point to directions for future inquiry.

The following is not suggested as a recipe for essay writing, but these points of guidance may be of some assistance in constructing a good paper.

In the *introduction*:

- State your aims or purpose clearly. What problem or issue are you discussing? Do not simply repeat or rephrase the question as this is one sure-fire way of putting any reader off your work.
- Make your conceptual framework clear. This gives the readers a basis for understanding the ideas which follow.
- Set your study in context. What is the significance of the topic?
- Outline the scope of your discussion (i.e. give the reader some idea of the spatial, temporal, and intellectual boundaries of your presentation). What case will you argue?
- Give readers some idea of the plan of your discussion—a sketch map of the intellectual journey they are about to undertake. Leave the reader in no doubt that your essay has a clear and logical structure (Burdess 1991, p. 127).
- Capture the reader's attention from the outset. Is there some unexpected or surprising angle to the essay? Alternatively, attention might be caught with relevant and interesting quotes, amazing facts, and anecdotes. Make your introduction clear and lively as first impressions are very important.

The best introductions are those which get to the point quickly and which capture the reader's attention (Bate & Sharpe 1990, p. 12).

In the *discussion*:

- Make your case. 'Who dunnit?'
- Provide the reader with reasons and evidence to support your views. Imagine your lecturer is sitting on your shoulder (an unpleasant thought!) saying 'prove that' or 'I don't believe you'. Disarm his/her scepticism.
- Present the argument logically, precisely, and in an orderly fashion.
- Accompany points of argument with carefully chosen, colourful, and correct examples and analogies.

In the *conclusion*:

- Can you resolve the research problem you set out to discuss? (Friedman & Steinberg 1989, p. 52). The conclusion ought to be the best possible answer to your essay question on the basis of the evidence that you have discussed in the main section of the paper. (Friedman & Steinberg 1989, p. 57). Bring ideas to fruition but do not repeat the main points or the introduction and do not introduce new material.
- If appropriate, discuss the broad implications of the work (Moxley 1992, p. 68).
- Tie the conclusion neatly together with the introduction. When you have finished writing your essay, read just the introduction and the conclusion. Do they make sense together? Finally, ask yourself: have I answered the question?
- Avoid cliched, phoney, mawkish conclusions (Northey & Knight 1992, p. 73) e.g. 'The tremendous amount of soil erosion in the valley dramatically highlights the awful plight of the poor farmers who for generations to come will suffer dreadfully from the loss of the very basis for their livelihood.'

In efforts to convey a sense of structure, and to make writing easier, some writers like to use *headings* throughout their essay. You are not compelled to use headings in essays—indeed, some lecturers actively discourage their use (check with your lecturer). However, for readers of an essay, headings do two things (Windschuttle & Elliott 1994, p. 109). First, they allow them to understand the overall structure of the essay rapidly and, at any stage, to know where they are in the essay's structure. Second, they make it easier to review particular sections to check a passage or point. Without headings, this rereading process may be difficult and slow. These two points provide a clue about the number and nature of headings that might be included in an essay. Include sufficient to provide a person quickly scanning the essay with a sense of the work's structure or intellectual 'trajectory'. To check this, write out the headings you propose to use. Is the list logical or confusing, sparse or detailed? Referring back to the essay itself, revise your list of headings until it provides a clear, succinct overview of your work.

Material relevant to topic

The material you present in your essay should be clearly and explicitly linked to the topic being discussed. To help clarify whether material is relevant or not, try the following exercise. When you have finished writing a draft of your essay, read each paragraph asking yourself two questions:

1 Does *all* of the information in this paragraph help answer the question?
2 *How* does this information help answer the question?

On the basis of your answers, edit. This should help you to eliminate the dross.

> Topic dealt with in depth

Have you simply slapped on a quick coat of paint or does your essay reflect preparation, undercoat and good final coats?! Have you explored all of the issues emerging from the topic? This does not mean that you should employ the 'shot-gun' technique of essay writing. Such poor essay-writing style sees the author indiscriminately put as much information as they can collect on a subject onto the pages of their essay.

Instead, be diligent and thoughtful in going about your research, taking care to check your institution's library, CD-ROMs, statistical holdings, other libraries, and information sources. Take notes/photocopies. Read. Read. Read. There is not really any simple way of working out whether you have dealt with a topic in sufficient depth. Perhaps all that can be said is that broad reading and discussions with your lecturer will provide some indications.

Quality of evidence

> Argument well supported by evidence and examples

One of the great temptations and weaknesses of student essays is the tendency to make unsupported generalisations (e.g. In these days when the divorce rate has never been higher ...; Cars in Australia are a leading cause of air pollution...) (Windschuttle & Windschuttle 1988, p. 154). Unsupported generalisations are indicators of laziness or sloppy scholarship and will usually draw criticism from essay markers.

Relevant examples, statistics, quotations from interviews and lectures, and other forms of evidence are required to support your case and to substantiate claims—to offer proof of your case or argument. In addition, most readers seek tangible examples which will bring to life or emphasise the importance of the points you are trying to make. Such information will be derived from good research (e.g. wide reading, interviews with appropriate people). Accordingly, in assessing essays many markers see the use of carefully selected and appropriate examples as useful indicators of diligence in research and the desirable ability to link concept/theory with reality.

Personal experience and observations may be incorporated as evidence in written work. For example, if you have spent several years as a police officer you may have some unique insights into an assignment on aspects of the geographies of crime and justice. Women and men who have spent time caring for children in new suburban areas may have valuable comments to make on the issue of social service provision to such areas. It is certainly valid to refer to your own experiences, but be sure to indicate in the text that it is to those that you are referring and provide the reader with some indication of the nature and extent of your relevant experience (e.g. 'In my 19 years as a police officer in central Melbourne...'). Where possible, support your personal observations with other sources, which readers may be able to consult.

Accurate presentation of evidence and examples

When you use examples, take care to ensure that they:

- are relevant
- are as up-to-date as possible
- are drawn from reputable sources (fully identified in your text with an appropriate referencing system)
- include no errors of fact.

Keep a tight rein on your examples. Use only those details you need to make your case.

Use of supplementary material

Effective use of figures and tables

Illustrations are used to make points more clearly, effectively or succinctly than they can be made in words. They are also usually more easily remembered than text. Histograms, pie charts, tables, schematic diagrams, and photos should supplement rather than duplicate text (Mullins 1977, p. 40).

When evaluating your use of illustrative material such as figures, tables, and maps, markers are interested in a variety of issues. They will check to see that you have referred to the illustration in your discussion and that the illustration makes the point intended. Assessors also look to see whether additional illustrative material might have been added to support the points you are making or to better organise the information you have presented. Illustrations do not need to come from your reading, but can be derived from your interpretations of material uncovered in your research (i.e. create your own illustrations where appropriate).

Illustrative material is very useful and is appreciated, but tables and graphics should be incorporated into the essay with a degree of care. Irrelevant graphics in great numbers do not impress any marker. Further, figures and tables should, in almost all instances, be located as close as possible to the text in which they are discussed. Unless you have a particularly good reason for doing so, do not put figures in an appendix at the end of the essay. Most readers find this very frustrating.

Illustrations effectively presented and correctly cited

Several types of illustrative material are commonly included in written work. These, and the accepted labels for each, are as follows:

Type of material	Label
Graphs, diagrams, and maps	Figure
Tables + word charts	Table
Photographs	Plate

Tables, figures, and plates can contribute substantially to the message being communicated in a piece of work. However, care must be taken in their presentation. Illustrative material should be:

- large (e.g. a figure can be expected to take up an entire page);
- comprehensible. Is the illustration easily understood and self-contained?
- legible;
- customised to your work. Do **not** submit an essay laden with marginally relevant photocopied tables and figures lifted directly from texts and journals. Where appropriate, redraw or rewrite the material to suit your aims;
- correctly identified with sequential arabic numerals beginning with 1. For example, you should not photocopy Table 17.1 from a text book and insert it in your essay as Table 17.1. Give it a number customised to your work (begin with Table 1 or Figure 1 etc.) and remove all trace of the photocopied numbering. When you place illustrative material in your work, ensure that tables/figures etc. are put in correct numerical order. For example, Figure 3 should precede Figure 4.

Titles of illustrative material should be clear and comprehensive. The title must fully specify the *subject* of the illustration, its *location*, and the *time period* to which it refers. For example, 'Vietnamese Born Population as Percent of Total Population, Adelaide Statistical Division, 1996' is a good title, whereas 'Vietnamese Population' is not. The *source* from which you derived the illustration should be specified. Failing to correctly identify the source is one of the most common problems associated with student use of illustrations in written work. The source should be acknowledged with an appropriate reference.

Maps and other diagrams should have a complete and comprehensive *key* or legend which allows readers to comprehend/decode the material shown. *Labelling* should also be neat, legible, and relevant to the message being conveyed by the illustration.

Refer to Chapter 4 for more information on the presentation of illustrative material.

Written expression and presentation

Fluent and succinct piece of writing

Unless you deliberately wish to obfuscate, there is little room for grandiloquence in effective written communication. In your essay you should be conveying knowledge and information, not showing how many big words you know (Booth 1985, p. 13). One straightforward means of checking that communication is clear is to ask yourself whether your writing would be understood by someone whose first language is not English. If you know someone in that position who is willing to proof-read your essay, give them a copy to look over. Another, very effective, means of checking the fluency of your writing involves putting a draft of your work away for several days and then reading it afresh. Odd constructions and poor expression which were not evident before will leap out to greet you. Yet another alternative is to make some mutual editing arrangements with friends. Make a deal. You will 'correct' their papers if they will 'correct' yours. Booth (1985, p. 6) notes that for over 2000 years it has been known that we see other people's mistakes more easily than we see our own.

Learning fluency in writing may seem impossible. There are, however, two fundamental secrets to success:

1 Take care to keep sentences short and as free of jargon as possible. Short sentences are easy to read. Short sentences convey ideas in a no-nonsense style.
2 Effective paragraphing is important. Although there are exceptions (see Clanchy & Ballard 1991, p. 37 for a discussion) paragraphs typically comprise three parts (Barrett 1982, p. 118):

- topic sentence—states the main idea (e.g. The depletion of Brazil's tropical rain forests is proceeding apace);
- supporting sentence(s)—why, how, examples to support the topic or to prove the point (e.g. There is little government action to end land clearance in fragile environments and private incentives to clear the land remain attractive);
- clincher—lets the reader know the paragraph is over. May summarise the paragraph, echo the topic sentence or ask a question (e.g. There seems to be little hope for the forests of the Amazon region).

Most paragraphs are unified by a *single* purpose or a single theme (Moxley 1992, p. 74). That is, a paragraph is a cohesive, self-contained expression of one idea. If your paragraph conveys a number of separate ideas, rethink its construction.

Paragraphs should relate to one another as well as to the overall thrust of the text. Get into the habit of using transitional sentences at the end of paragraphs to carry the reader onto the next paragraph. Bate and Sharpe (1990) discuss effective paragraphing in some detail.

Make every word count. Waffle is easily detected and it makes assessors suspect that you have little of substance to say. Prune unnecessary words and phrases from your work. Remember, the objective in an essay is to answer the question or to convey a body of information—not to write a specified number of words!

Grammatical sentences

One simple way of detecting difficulties with grammar is to have that trusty friend read your essay out loud to you. If that person has difficulty and stumbles over sentence constructions, it is likely that the grammar is in need of repair. Another simple means of avoiding problems is to keep sentences short and simple. Not only are long, convoluted sentences often difficult to understand, they are also grammatical minefields.

Correct punctuation

Check the material in Chapter 8 for a review of common punctuation problems. Take particular care with the use of apostrophe! If there is anything you do not understand, ask your lecturer.

Correct spelling throughout

Poor spelling has the capacity to bring even the best of work into question. Spelling errors emerge repeatedly as a problem in university-level essays. In these days when many papers are written with the help of a word-processing package, there is little excuse for incorrect spelling. If you write your essay on a computer with a word-processing package be sure to use the spelling checker before you submit the essay for assessment but remember, distinctions between words such as there/their, too/two/to, and bough/bow will not show up. If you write by hand or use a typewriter, go back to that friend you have been troubling for assistance. Have them look over your paper with an eye to the spelling and grammatical errors.

Legible, well set out work

> Poor presentation can prejudice your case by leading the reader to assume sloppiness of thought. (Bate & Sharpe 1990, p. 38)

Essays that are difficult to read because of poor handwriting can infuriate assessors. Frustrations emerge because it is very difficult to maintain a sense of your argument if reading must be repeatedly interrupted to decipher individual words. Wherever possible use a typewriter or word-processor to produce the final copy of your paper.

To allow room for the assessor's comments, all work should be double-spaced and have a large left-hand margin (if in doubt, try 3.5–4 cm). Print your assignment on only one side of A4 pages.

Nicely presented work suggests pride of authorship. You are likely to find that presentation does make a difference—to your own view of your work as well as to the view of the assessor.

Reasonable length

A key to good communication is being able to convey a message with economy (consider the communicative power of some short poems). Take care not to write more words than have been asked for. Most people marking essays do not want to read any more words than they have to. How would you feel about a 2500 word essay submitted with 60–100 others when the imposed word limit was 2000 words? Similarly, avoid masking a scarcity of ideas with verbose expression. If you find you do not have enough to say in an essay, perhaps you need to do some more research—not writing.

Sources/referencing

An introductory word of advice on references: when writing an essay, be sure to insert citations as you go along. It is very difficult to come back to a paper and try to insert the correct references (Hodge 1994).

Adequate number of sources

The quality of evidence you use in your work will be considered very carefully by markers. You are expected to demonstrate that you have conducted extensive research appropriate to the topic and to the level of the course you are doing (e.g. first year essays might draw from secondary sources such as books and journals; third year research essays might require library research, interviews, fieldwork, and information derived from other primary information sources).

You will be expected to draw your evidence from, and substantiate claims with, *up-to-date*, *relevant* and *reputable* sources (e.g. in most instances a magazine such as the *National Enquirer* would not be considered reputable). Unsubstantiated and vague claims have no place in university work. So, for example, instead of saying that 'there is lots of air pollution in Mexico City', you would be expected to both elaborate on what you mean by 'lots' and to let the reader know the basis for your authority to make that claim. Thus, the claim might be modified: 'The United States Environmental Protection Agency (1993, p. 369) notes that Mexico City has the highest levels of air pollution in the world'.

Take care not to place heavy reliance on a small number of references. Most assessors give some weight to the *number and range of references* you have used for your work. This is to ensure that you have established the soundness of your case by considering evidence from a broad range of possible sources. For example, one might be suspicious of an essay examining the consequences of hospital privatisation for Australian rural health care delivery if the essay was based largely on documents produced only by the Liberal Party or by the Labor Party.

Adequate acknowledgement of sources

Ideas, facts, and quotations *must* be attributed to the source from which they were derived. Failure to acknowledge sources remains a common and potentially dangerous error in student essays. Serious omissions may constitute plagiarism (see the notes on plagiarism in Chapter 8). Acknowledgements should be made using an appropriate system of referencing.

Now, all of this business of acknowledging sources may seem to you to be a painful waste of time. But, like most things, there are reasons for it. People writing in an academic environment acknowledge the work of others for two main reasons:

- to attribute credit (and sometimes blame) for the acknowledged author's contribution to knowledge; and
- to allow interested readers the opportunity to pursue a line of inquiry should they be stimulated by something that has been cited. For example, in an essay on the economic geography of South Australian brewing, someone might write: 'Jones (1994, p. 16) observes that there is a great deal of money to be made by private investors by investing in the brewing business'. 'Aha' you think, 'I can make some quick dollars here. All I need to do is find Jones' book and read about how this might be done'. Thanks to the appropriate acknowledgement of Jones' idea in the essay you have read, you can rush to the nearest library and track down Jones' words of wisdom on money-making. Lo and behold, instant millionaire!

Quite often, people new to essay writing save all the acknowledgements of references contained in a paragraph until the end of that paragraph. There they place a string of names, dates, and page numbers. This is incorrect and annoying because the reader has no way of establishing which ideas/concepts/facts are being attributed to whom. References should be placed as close as possible to the ideas or illustrations to which they are connected.

Correct and consistent in-text referencing style

Full and correct acknowledgement of the sources from which you have derived quotations, ideas, and evidence is a fundamental part of the academic enterprise. For some reason, referencing problems remain amongst the most frequently encountered difficulties in undergraduate written work. Acknowledging the contribution of others to the essay you have written should not be difficult if you follow the instructions on referencing provided in Chapter 8.

Reference list correctly presented

The most common—and most easily rectified—problems in essay writing emerge from incorrect acknowledgement of sources. Repeatedly in students' essays, referencing is done improperly and reference lists are formatted incorrectly. Many of those people who mark essays consider that problems with the relatively simple matter of referencing reflect more serious shortcomings in the work they are reading. In consequence, it is advisable to follow carefully the instructions on referencing. If you do not understand how to refer to texts in an essay, see your lecturer.

Aside from the mechanical problem of setting up a reference list, it is clear that there is a related editing problem in many essays. Students often refer to a source within the text of their essay but do not provide the full bibliographic details of that source in their list of *References Cited* (at the end of the essay). Before you submit an essay for assessment, be sure that all in-text references have a corresponding entry in the list of *References Cited*. Further, in most cases, the list of *References Cited* should include *only* those references you have actually cited in the paper.

Demonstrated level of individual scholarship

'Scholarship' is one of the most important and perhaps the least tangible of the qualities that make a good essay. Above all, the essay should clearly be a product of *your* mind, of *your* logical thought. The marker will react less than favourably to an essay which is merely a compilation of the work of other writers. In considering matters of scholarship, then, essay markers are searching for judicious use of reference material combined with *your individual insights*.

Whilst scholarship requires that you draw from the work of other writers, you must do so with discretion. Use direct quotations sparingly (if you do use quotations, be sure to integrate them with the rest of your text). Keep paraphrasing to a minimum.

You might argue that novice status in the discipline means that you must rely heavily on other people's work. Obviously you might encounter problems when you are asked to write an essay on a subject that, until a few weeks ago, may have been quite foreign to you. But do not be misled into believing that in writing an essay you must produce some earth-shattering exposition on the topic you have been assigned. Instead, your assessor is looking for evidence that you have read on the subject, *interpreted* that reading, and outlined a reasonable argument/statement, based on that interpretation, which satisfactorily addresses the essay topic.

Essay Assessment Schedule

Student Name: Grade: Assessed by:

The following is an itemised rating scale of various aspects of written assignment performance. Sections left blank are not relevant to the attached assignment. Some aspects are more important than others, so there is no formula connecting the scatter of ticks with the final grade for the assignment. Ticks in either of the two boxes left of centre means that the statement is true to a greater (outer left) or lesser (inner left) extent. The same principle applies to the right-hand boxes. If you have any questions about the individual scales, comment, final grade or other aspects of this assignment, please see the assessor indicated above.

Quality of argument

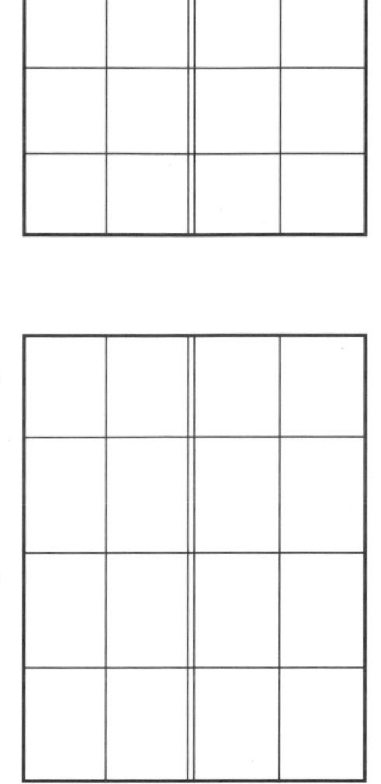

The argument fully addresses the question	The argument fails to address the question
Logically developed argument	Writing rambles and lacks logical continuity
Writing well structured through introduction, body and conclusion	Writing poorly structured, lacking introduction, cohesive paragraphing and/or conclusion
Material relevant to topic	Much material is not relevant
Topic dealt with in depth	Superficial treatment of topic

Quality of evidence

Argument well supported by evidence and examples	Inadequate supporting evidence or examples
Accurate presentation of evidence and examples	Much evidence incomplete or questionable
Effective use of figures and tables	Figures and tables little used or not used when needed
Illustrations effectively presented and correctly cited	Illustrations poorly presented or incorrectly cited

Written expression and presentation

Fluent and succinct piece of writing					Clumsily written, verbose, repetitive
Grammatical sentences					Many ungrammatical sentences
Correct punctuation					Much incorrect punctuation
Correct spelling throughout					Much incorrect spelling
Legible, well set out work					Untidy and difficult to read
Reasonable length					Over/under length

Sources/Referencing

Adequate number of references					Inadequate number of references
Adequate acknowledgement of sources					Inadequate acknowledgement of sources
Correct and consistent in-text referencing style					Incorrect and inconsistent in-text referencing style
Reference list correctly presented					Errors and inconsistencies in reference list

Demonstrated level of individual scholarship

High					Low

Assessor's comments

2

Writing research reports and laboratory reports

Your lecturer has assigned you a research project and, as if that was not difficult enough, you have been asked to report on your work—in writing. This chapter about writing research reports and laboratory reports is intended to help you complete that task. Indeed, by providing some guidance on how to write a particular kind of report, the following pages might also help you to undertake the research.

Why Write a Report?

There are at least three good reasons for learning to write good research reports. These range from the practical to the principled.

First, there is a *vocational claim*. Academic and professional writing often involves the communication of research findings (Friedman & Steinberg 1989, p. 24). Urban planners, market researchers, academics, environmental scientists, and intelligence analysts can all expect to undertake research and to write associated reports in the course of their employment. Indeed, getting and maintaining employment in areas related to geography and the environment often requires the effective conduct and communication of research. That communication is usually to an audience that anticipates answers to a certain set of questions that must nearly always be answered, irrespective of the character of the project. Consequently, it is important to be familiar with the ways in which research results are customarily conveyed from one person to another (i.e. the conventions of research communication).

Second, research reports and papers are a *fundamental building block of knowledge*. Each report is the final product of a process of inquiry. Through the communication of research findings we contribute to the development of practically adequate understandings of the ways in which the world works. 'Practically adequate' means those understandings will not necessarily be absolutely and forever right. Instead, they work and make sense here and now. Some event or discovery may see them change tomorrow.

Third, there is a *moral responsibility to present our research honestly and accurately*. Through our research writing we help to forge understandings about the ways in which the world works. Representing the world to other people in ways that we understand is to play an enormously powerful role. To a degree, people entrust us with the creation of knowledge. Given that trust, our actions must be beyond reproach. In part acknowledgement of that provision of trust we must provide peers, colleagues, and interested observers with accurate representations of our actions. Therein lies a critical role of research reports and a most important reason for writing them well.

What are Report Readers Looking For?

Reports answer five classic investigative questions (Eisenberg 1992, p. 276).

- What did you do?
- Why did you do it?
- How did you do it?
- What did you find out?
- What do the findings mean?

The person reading or marking your report seeks clear and accurate answers to these questions. Because reports are sometimes long and complex, the reader will also appreciate some help in navigating their way through the document (Windschuttle & Elliott 1994, p. 261). Make the report clear and easy to follow through easily understood language, a well-written introduction, suitable headings and subheadings and, if appropriate, a comprehensive table of contents.

Some forms of report, especially lab reports, will answer the five investigative questions through a highly structured progression (e.g. introduction, methods, results, discussion) written in a way which would allow another researcher to repeat the work. For example, an environmental scientist reviewing present-day salinity levels in Australia's River Murray, or a demographer conducting a statistical study on the use of contraceptive measures in New Zealand, is likely to conduct the study as impartially as possible and to record their research procedures in sufficient detail to allow someone else to reproduce the study. For such forms of inquiry, repetition is an important means of verifying results. (This notion of reproducibility or replication is discussed fully in Sayer 1992.)

Other reports, such as those on studies involving qualitative research methods (e.g. interviews, participant observation, textual analysis) will usually answer the five questions identified above, but in a rather more literary, politicised, and less formally structured form. Because a good deal of qualitative research cannot be repeated exactly (e.g. interviewees and texts may not be available to all researchers) the most appropriate form of verification is corroboration or, in other words, substantiation or confirmation. Other researchers simply cannot repeat the study exactly as you conducted it and so there is less emphasis placed on those parts of the report which might allow the study to be repeated. In addition, the boundaries between results and discussion may become blurred. This is not intended to imply any research sloppiness. In a report drawn from qualitative research, it is still necessary to outline accurately the means by which information was collected, how it was gathered, and the processes involved in its interpretation.

To make the points about corroboration and writing a little clearer, consider the way a trial for murder works. We cannot expect the murder to be repeated in order to establish who the murderer really was! Instead, lawyers and police assemble all manner of evidence to support or refute a claim that some particular individual committed the atrocity. Evidence is gathered and admitted to the case according to established rules of evidence. Police and lawyers then tell a story (or conflicting stories) about an event that cannot be reproduced and base their tale(s) on a variety of interlinked, mutually supporting evidence.

Now, geographers reporting on the social construction of an Australian city, or about gay men's perceptions of everyday places, will use different procedures and will write their research in different ways from coastal geomorphologists studying sand-grain size and longshore drift or economic geographers writing about the demographic characteristics of Australian country towns. Yet they will still usually answer the five basic investigative questions identified above, even though style and emphases may make the reports quite different in presentation.

The great diversity of research topics found within geography and environmental science means that you are likely to be asked to write research reports of different types throughout your degree. Reflecting that potential diversity, this chapter provides an introduction to report writing *in general*. Some specific references are made to laboratory writing. You may find it useful to look over a few published reports as examples of the sort of work that might be produced (e.g. Baker, Robertson & Sloan 1993, Coventry et al. 1993, Bell 1995) or to ask your lecturers if they have written reports which you might consult as examples.

Report Writing—General Layout

It should be clear from the paragraphs above that although research and laboratory reports will usually answer Eisenberg's five investigative questions, there is no single correct research report style. The best way to organise a research report is determined by the type of research being carried out, the character and aims of the author, and the audience to which the report is directed (discussed in more detail in Mohan, McGregor & Strano 1992, pp. 220–222). Accordingly, the following guidelines for report writing cannot offer you a recipe for a 'perfect' research report. Good reports include well executed research and the will to communicate the results of your work effectively.

Having acknowledged that there is no single 'correct' report writing style, it is fair to say that over time a common pattern of report presentation has emerged. That pattern reflects a strategy for answering the five investigative questions identified above. Through repeated use, it is also a structure of presentation many readers will expect to see in a report. If you are new to report writing, and unless you have been advised otherwise, it may be useful to follow the general pattern outlined below. If you are more experienced and believe there is a more effective way of communicating the results of your work, try out your own strategy. Remember, however, that you are guiding the reader through the work: you will have to let your audience know if you are doing anything they might not expect.

Short research reports and laboratory reports generally comprise a minimum of six sections, four of which deal specifically with the investigative questions identified above:

- Abstract/Executive summary
- Introduction (what you did and why)
- Materials and methods (how you did it)
- Results (what you found out)

- Discussion (what the results mean)
- References.

Longer reports may add:

- Title page
- Letter of transmittal
- Acknowledgements
- Table of contents
- Recommendations
- Appendices.

Although these headings point to an order of report presentation, there is no need to write the sections in any particular sequence. Indeed, you may find it useful to follow Woodford's (in Booth 1993, p. 2) advice to label several sheets of paper headed Title, Summary, Introduction, Methods, Results etc. and use these to jot down notes as you work through the project. Then begin your report by writing the easiest section (the methods section in many cases).

The following pages outline the form and function of the common components of research reports. Discussion also elaborates on some of the issues that contribute most significantly to effective research presentations in an academic setting. Those same issues form the basis of an assessment schedule for research reports and laboratory reports.

Preliminary material

Title page

The main function of the title is to contribute towards the overall mission of communicating information effectively (Gray 1970, p. 70). The best titles are usually short, accurate, and attractive to potential readers. Use a subtitle if a fuller description is required. The title should provide instant identification of the content (Mohan, McGregor & Strano 1992, p. 226). When you have finished writing your report, check that the title matches the results and discussion. This is important, as it is the title which gives readers a strong sense of the purpose of your work. You need to ensure that the impression gained matches the content of the report.

An example of a functional and informative title is: 'Social consequences of homelessness for men in Adelaide, South Australia (1990–1995)'. This title lets the reader know the topic, place, and time period. An example of a bad title on the same subject matter is: 'Men and homelessness'.

In a discussion of technical reports Gray (1970, p. 74) observes that the title page of a report should also include:

- your name, position, and organisational affiliation
- name of the person and/or organisation to whom the report is being submitted
- date of issue of the report.

Modify these recommendations to suit the academic setting in which you find yourself. For example, date of issuance might be the due date or the date you submit the assignment for assessment. The person to whom the report is submitted may be your lecturer.

Letter of transmittal

Reflecting the fact that reports are often commissioned, a letter of transmittal is sometimes included. This letter personalises the report for the reader who commissioned (or asked for) the report and lets them know what parts of it may be of particular importance to them (Mohan, McGregor & Strano 1992, p. 227).

Abstract/Executive summary

Of all sections of the report other than the title, it is this which is the most likely to be read. It is important therefore to make it easy to read and understand. A good abstract outlines:

- objectives
- methods
- results
- principal conclusions of a research report.

An abstract is a coherent and concise statement, intelligible on its own, and written in introduction–body–conclusion form. Abstracts are *not* written in the form of notes. Abstracts are limited in length (typically 100–200 words) and are designed to be read by people who may not have the time to read the whole report. For that reason, do not write an abstract as if it is the alluring back cover of a mystery novel. Let your reader know what your research is about; do not leave them in suspense.

All information contained in the abstract must be discussed in the main part of the report (Behrendorff 1995, p. iii). Abstracts are located at the beginning of reports, although they are usually the last section written.

Acknowledgements

If you have received valuable assistance and support from some people or organisations in the preparation of the report they should be acknowledged. As a general rule thank those people who genuinely helped with aspects of the work, such as proof reading, preparing figures and tables, solving statistical or computing problems, taking photos, or doing the typing.

Table of contents

This is required in longer works to assist the reader in following the structure of the report. It should accurately and fully list *all* headings and subheadings used in the report with their associated page numbers. The table of contents occupies its own page and must be organised carefully with appropriate spacing. Make sure that the numbering system used in the table of contents is the same as that used in the body of the report.

Included after the table of contents, and on separate pages, are a list of figures and a list of tables. Each of these lists contains, for each figure or table, its number, title, and the page on which it is located.

Introduction—why did you do this study?

The introduction of a report answers the following questions:

- What do you hope to learn from this research?
- What question is being asked?
- Why is this research important? (social, personal, and disciplinary significance)

When you write your introduction imagine that readers are unfamiliar with your work and that they really do not care about it. Sell your work. Let your audience know why this report is important and exactly what it is about. When the reader knows these things they will be better able to grasp the significance of the material you present in the remainder of the report.

Commonly, a literature review will be required for any discussion of the significance of the problem. (The literature review is sometimes presented as a separate part of the report, after the introduction and before the discussion of materials and methods.) As its name suggests, it is a comprehensive, but pithy, summary of publications and reports related to your research. The review is intended to provide the reader with an understanding of the conceptual and disciplinary origins of your study. The literature review also provides you with an opportunity to discuss the ways in which your study may contribute to, fit in with, or differ from available work on the subject.

> The literature review must not be a series of quotations or a series of findings of other authors' work, simply strung together. It should consist of a critical analysis of previous work with the aim of leading the reader to the point where it becomes clear why the project has been proposed. (Behrendorff 1995, p. 4)

A good literature review will normally discuss significant books written in the field (e.g. if you are writing about town locations, you would probably refer to the landmark works of Christaller and Lösch on central place theory), notable books and articles on the broad subject in the past four or five years, and all available material on your narrow, specific research area.

When your reader has finished reading your introduction, they should know exactly what the study is about, what you hope to achieve from it, and why it is of significance. If they have also been inspired to read the remainder of the report, so much the better!

Materials and methods—how did you do this study?

Once you have told your reader something about the background to the study it is then necessary to let them know how you did the research. Many readers will be trying to confirm that you have chosen the methods most appropriate to the topic. Provide a precise and concise account of the materials and methods used to conduct the study and why you chose them. Let your reader know exactly how you did the study and from where you got your data. Try not to leave out essential details. For example: 'If you used "alcohol", in some part of a physical geography experiment, say which alcohol. If you controlled or even measured the humidity and ventilation..., say so: they may be nearly as important as temperature.' (Booth 1993, p. 5).

A good description of materials and methods is one which would enable the reader to duplicate the investigative procedure if they had no source of information about your study other than your report. Note, however, that in qualitative studies duplication of procedure is unlikely to lead to identical results. Instead, the outcome may be results which corroborate, substantiate or, indeed, refute, those achieved in the initial study (see Sayer 1992 for a full discussion).

The methods section of a report comprises up to three parts (Dane 1990, pp. 219–21) depending on the specific character of the research. The parts may be written and presented as a single section or under separate subheadings.

Sampling

An important part of the materials and methods section of a report is a statement of *how* and *why* you chose some particular place, group of people or object to be the focus of your study. For example, if your research concerns people's fears and the implications those fears have for the use of urban space, why have you chosen to confine your study to some specific suburb of a specific Australian city? Having limited the study to that location, why and how did you choose a small group of people to speak to from the much larger total local population? Alternatively, in an examination of avalanche hazards in New Zealand's South Island high country, why and on what bases did you limit your study to those risks associated with one popular ski area?

In the sampling section of your report, your reader will appreciate answers to the following questions:

- *Who/what* specific group, place or object have you chosen to study? You may have already stated in the introduction that you were exploring the attitudes of Papua New Guinean women to birth control, but you now need to identify the specific group and number of women you are going to interview or to whom you will administer questionnaires (e.g. 3000 urban dwelling women of child-bearing age). Or, in your study about supernatural explanations of unusual landscape features, you might have determined that you will limit your study to Incan constructions in Peru within a 100 kilometre radius of the historically important town of Cuzco.
- *Why* did you make that choice? Why did you limit the study to 3000 Papua New Guinean women of child-bearing age and not to a smaller group of rural dwelling women? Or in our other example, why Peru and not Easter Island? Why 100 kilometres? Why Cuzco and not Machu Pichu or Aguas Calientes?
- *How* did you select the unit(s) of study? That is to say, what specific sampling technique did you employ (e.g. snowball, simple random, typical case, cluster area, random traverse)? There is no

need to go into great detail about the technique, such as describing any computer programs used in your sampling, unless the procedure was unusual.
- What are the *limitations* and shortcomings of the data or sources?

Of course, if you are reporting a field study, a general description of the study site is needed. Do not forget to include a map, for it may save you a great deal of writing and will almost certainly provide you reader with a clearer sense of the place you are describing than might the proverbial thousand words. Photographs may also be helpful. If the site description is especially lengthy, create a separate section for it (Brower, Zar & von Ende 1990, p. 22).

Apparatus or materials

Provide a brief description of any special equipment or materials used in your study. For example, briefly describe any questionnaires used in your survey (also attach a copy as an appendix) or identify and, where necessary, describe survey equipment used in field observations. Again, do not hesitate to use figures and plates if it seems they might be useful to your reader.

Procedure

This section contains specific details about how the data were collected and the methods used to interpret the findings. If your study was based on a questionnaire survey, tell the reader about the process of questionnaire administration. When were the forms handed out? Who administered the surveys? What happened when targeted individuals could not be contacted? Enough detail should be provided to allow the reader to replicate your procedures. How, when and where did you conduct your research? How did you describe the study to participants? While it is important to describe the way in which data were collected and how it was interpreted, you can assume that the reader has some degree of competence in the use of the method discussed (unless it is particularly novel). There is no need to explain, for example, how you calculated the standard deviation of Perth's annual rainfall. It is important however that you *justify* your selection of procedure. Why did you choose one method over others? What are the advantages and disadvantages and how did you overcome any problems you encountered?

Results—what did you find out?

The results section of a research report is typically a dispassionate, factual account of findings. It outlines what occurred or what was observed. In laboratory reports, most physical geography/environmental reports, and in empirically-based human studies, it is *not* customary practice to present conclusions and interpretations in the results section. However, in some qualitative studies it is considered appropriate to combine the results with an interpretive discussion.

A key to effective presentation of results is making them *as comprehensible to your readers as possible*. Towards this end it may be appropriate to provide your reader with an overview of what you are going to do in this section. Use maps, tables, figures, and written statements creatively to convey key information emerging from the study. Try to avoid presenting listings of raw data in the results section. That sort of detailed information is better placed in an appendix. Instead, summarise your numerical data using graphic devices such as line graphs and statistical methods such as means, frequency tables, and correlation coefficients.

The results section will often contain a series of subheadings. These usually reflect subdivisions within the material being discussed, but sometimes reflect matters of method. In general, however, try to avoid splitting up the results section on the basis of methods, since it may suggest that you are 'allowing the methods rather than the issues to shape the problem' (Hodge 1994, p. 2).

If you have not already done so in the report, the results section is an appropriate place in which to identify the limits of your data.

Discussion and conclusion—what do the findings mean?

> I am appalled by...papers that describe most minutely what experiments were done, and how, but with no hint of why, or what they mean. Cast thy data upon the waters, the authors seem to think, and they will come back interpreted. (Woodford 1967, p. 744)

The discussion is the heart of the report. Perhaps not surprisingly, it is also the part which is most difficult to write and, after the title, abstract, and introduction, is the section most likely to be read thoroughly by your audience. Readers and assessors will be looking to see if your work has achieved its stated objectives. So, take particular care when you are writing this part of your report.

The discussion has two fundamental aims:

- to explain the results of your study. Why do you think the patterns or lack of patterns you uncovered emerged?
- to explore the significance of the study's findings. What do the findings mean for the discipline, for humanity, for you? Because research is a shared enterprise, with people working in mutually

supportive ways to investigate issues and problems, it is very important to embed your findings in their larger academic, social, and environmental contexts. Consider and make explicit the ways in which your work fits in with that conducted by other people and the degree to which it might have broader importance. David Hodge makes the point:

> Remember that research should never stand alone. It has its foundations in the work of others and, similarly, it should be part of what others do in future. Help the reader make those connections. (Hodge 1994, p. 3)

On a final note, the concluding sections of the report might also offer suggestions about improvements or variations to the investigative procedure that might be useful for further work in the field. Where do we go from here? Are there other methods or data sets which should be explored? Has the study raised new sets of questions? (Hodge 1994, p. 3).

Recommendations

If your report has led you to a position where it is appropriate to suggest particular courses of action or solutions to problems, you may wish to add a recommendations section. This could be included within the conclusion or accorded a free-standing place. In some reports, recommendations are placed at the front, following the title page (Gray 1970, p. 6). Recommendations should be based on material covered in the report (Mohan, McGregor & Strano 1992, p. 228).

References

For information on citing references in a research report, see Chapter 8.

Appendices

Material which is not essential to the report's main argument (Gray 1970, p. 49) and is too long or too detailed to be included in the main body of the report is placed in an appendix at the end of the report. For example, you might include a copy of the questionnaire you used or background information on your study area or pertinent data which is too detailed for inclusion in the main text. However, be warned, your appendix should *not* be a place to put *everything* you collected in relation to your research but for which there was no place in your report (Kane 1991, p. 187). Appendices are usually located after the conclusions but before the references.

Written expression and presentation

Language of the report

Many report writers try to hide the impossibility of objectivity in research behind a facade of unemotive terms and the language of the third person (e.g. 'it was considered' rather than 'I considered'). The public pressures to write in such a language remain quite intense. People whose educational histories were wrought in physical science backgrounds or in an earlier phase of social research, and who now hold positions of influence over the ways in which research is presented and accepted, often distrust or dismiss reports that acknowledge through their writing style the simple subjectivity of all research endeavours. A report written in the first person or employing emotive terms that clearly reflect the views of the author may be considered less scientific and less valuable than a piece of work reporting similar results in the more commonly accepted, yet fundamentally dishonest[1] language of claimed objectivity. So, when you write a research report, be aware of these politics of rhetoric.

There is no law of writing that one must be impartial and use a particular style of language. However, what you should do is *remember who your audience is and what conventions of presentation they expect*! Imagine, for example, you have slightly left-wing political leanings and have been commissioned to write a potentially influential report for a slightly right-wing government on the consequences for rural areas of public hospital closures. How would you present 'yourself' in that report? Or perhaps more accurately, how are you likely to 'hide' in that report? Rather differently, in a physical geography assignment on dryland salinity, your lecturer is likely to expect a dispassionate representation of your activities and results. That means a particular way of writing. If you have any questions or concerns about whether to include or exclude yourself from your report writing, ask your lecturer.

1 By no means do I mean to suggest here that those individuals who write 'scientific' reports are themselves dishonest or otherwise immoral. Instead, I would argue that attempts to 'objectify' research writing repeatedly renounce the fundamental humanity of (social) scientific inquiry. The consequent social structural construction is one in which a dehumanised research process has become the most commonly accepted 'valid' form of inquiry. Research is given validity through its presentation as having emerged from some unquestionable extra-human authority and not from the works of a single researcher or group of researchers. The social construction of science, paradigmatic contexts of understanding and, consequently, the practically adequate, temporary nature of the ways we understand the world are all denied.

Another matter of language which needs to be emphasised is that of jargon. The word jargon has two meanings. Most commonly, jargon refers to technical terms used inappropriately or when clearer terms would suffice: in other words, gibberish. Less commonly, it means words or a mode of language intelligible only to a group of experts in the field (Friedman & Steinberg 1989, p. 30). There will be occasions in report writing when you will find it necessary to use jargon in the second sense of the word. You should never be guilty of using words in the more common, first sense of the term jargon. Remember, you are writing to communicate ideas to the intended audience as clearly as possible. Use the language which allows you to do that. KISS (i.e. Keep It Simple Stupid) your audience.

Presentation

When you write a report, it is important to match the high quality of your research with high quality presentation. Your reader will not only be impressed by high quality presentation, they may also find it easier to understand your results. Be sure that the report is set out in an attractive and easily understood style. Care taken in presentation will also suggest to people that you have taken care in preparation. *Care* in presentation is to be emphasised here, not gaudiness and decoration. Do not spend a fortune at the printing shop getting your report lettered in gold leaf, for instance. People tend to be suspicious of overly decorated reports and in a professional environment (e.g. consulting) may also question the costs of production.

Instead, get the fundamental matters straight. For instance:

- Use the same size paper throughout. Try to avoid, for example, having the report and some of the appendices on A4 paper and the remainder of appendices on legal sized paper or A3. There are some occasions, however, when use of same size sheets is impractical (e.g. inclusion of maps).
- Number all the pages.
- Use SI (*Systeme Internationale*) units (i.e. metric) in describing measures.

Together, these matters of written expression and presentation may have an important influence on the overall effectiveness of your report.

Writing a Laboratory Report

A laboratory report might be imagined to be a particular form of research report and, hence, the advice provided earlier in this chapter is applicable. You are strongly advised to refer to those pages before you begin writing a report on an experiment or other laboratory activity.

Typically, a lab report dispassionately and accurately recounts experimental research procedures and results. A good report is written so that another student-researcher could repeat the experiment in exactly the same way as you did (assuming, of course, that you employed correct procedures) and could compare their results with yours. Clearly then, you must be both meticulous in outlining methods and accurate in your presentation of results. A note of caution—being meticulous does not mean that you have to be tedious in your account of methods and results. Try not to overdo the detail. In the context of the specific experiment that you are doing, record those things which are important. What did *you* need to know to do the experiment? Let your reader know that. One could almost argue that the best 'Materials and methods' sections in laboratory reports ought to be written by novices who are less likely to make assumptions about readers' understanding of experimental methods than are more experienced researchers.

By convention, laboratory reports follow the order of research report presentation outlined earlier in this chapter, that is:

- Title page
- Abstract
- Introduction
- Materials and methods
- Results
- Discussion
- References
- Appendices

Depending on the specific nature of your experiment, your lab supervisor may not require all these sections. For example, if your report is simply an account of work you undertook during a single class laboratory time it is possible that there will be no need for you to include an abstract, references, and appendices. On the matter of references, your lecturer may be impressed if you take the care and attention to relate your day's lab work to appropriate reference material. Clearly, though, if your lab work is conducted over a longer period than a single class, you will have the opportunity to consult pertinent reference materials.

Research Report Assessment Schedule

Student Name: Grade: Assessed by:

The following is an itemised rating scale of various aspects of research report performance. Sections left blank are not relevant to the attached assignment. Some aspects are more important than others, so there is no formula connecting the scatter of ticks with the final grade for the assignment. Ticks in either of the two boxes left of centre means that the statement is true to a greater (outer left) or lesser (inner left) extent. The same principle applied to the right-hand boxes means that the topic was treated somewhat superficially in the assignment. If you have any questions about the individual scales, comment, final grade or other aspects of this assignment, please see the assessor indicated above.

Purpose and significance

Statement of problem or purpose is clear and unambiguous					Statement of problem or purpose is unclear or ambiguous
Research objectives outlined precisely					Research objectives unclear
Disciplinary, social and personal significance of the research problem made clear					Problem not set in any context
Documentation fully outlines the evolution of the research problem from previous findings					No reference to earlier works or incorrect references

Description of method

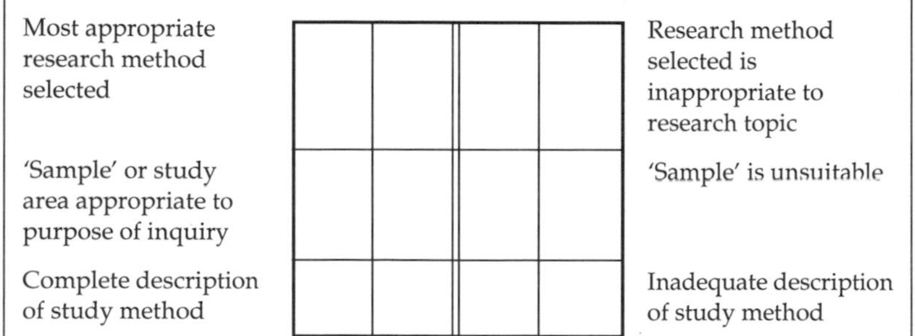

Most appropriate research method selected					Research method selected is inappropriate to research topic
'Sample' or study area appropriate to purpose of inquiry					'Sample' is unsuitable
Complete description of study method					Inadequate description of study method

Quality of results

Evidence of extensive primary research					Little or no evidence of primary research
Limitations of sources made clear					Inappropriate sources accepted without question
Relevant results presented in appropriate level of detail					Relevant results omitted or suppressed

Discussion & interpretation

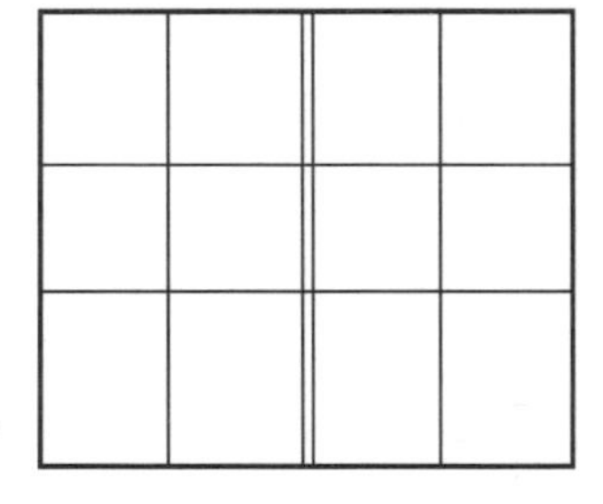

No errors of interpretation (e.g. logic, calculation) detected					Many errors of interpretation
Any limitations of findings made clear					Limitations of findings not identified
Discussion connects findings with relevant literature					No connection between findings and other works

Conclusions

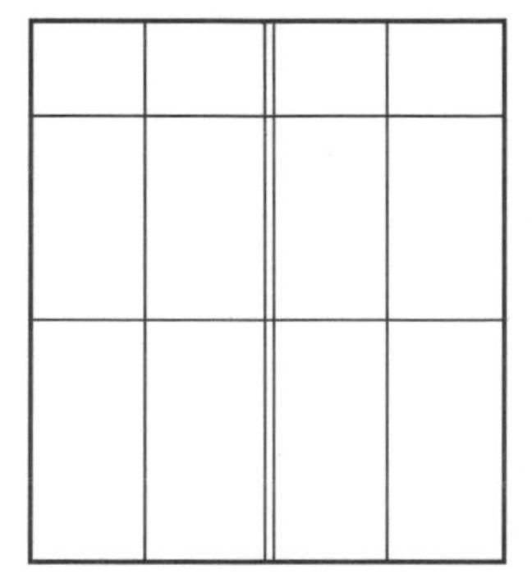

Significance of findings made clear					No significance identified
Conclusions based on evidence					Little or no connection between presented evidence and conclusions
Stated purpose of research achieved					Little or no contribution to solution of problem or achievement of purpose

Use of supplementary material

Effective use of figures and tables					Figures and tables little used or not used when needed
Illustrations effectively presented and correctly cited					Illustrations poorly presented and/or incorrectly cited
Detailed statistical analyses and tables placed in appendices					Excessively detailed findings in text

Written expression and presentation

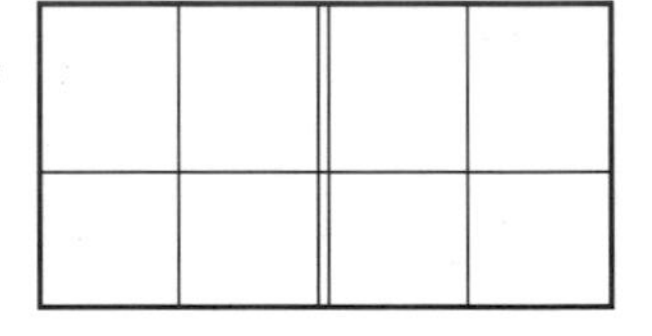

Document follows assigned report format					Little or no adherence to report presentation conventions
Clearly and correctly written					Poor written expression

Sources/referencing

Adequate number of sources					Inadequate number of sources
Adequate acknowledgement of sources					Inadequate acknowledgement of sources
Correct and consistent in-text referencing style					Incorrect or inconsistent referencing style
Reference list correctly presented					Errors and inconsistencies in reference list

Assessor's general comments

3

Writing reviews, summaries, and annotated bibliographies

One important part of academic endeavour is making sense of the works of other people. This chapter provides some advice on writing book and article reviews, summaries, and annotated bibliographies, exercises which specifically require the ability of comprehension. The discussion is structured around the sorts of matters readers and assessors of each form of work typically seek.

Why Write a Review?

You have just been asked to write a review. There are a number of reasons your lecturer may have asked you to do this:

- to familiarise you with a significant piece of work in the field (the assigned text to review);
- to consider the place of that text within the discipline (how does it relate to other works in the field?);
- to practise your capacity for critical thought.

You should also be aware of the specific functions of a review. Reviews serve an important role in the professional and academic world. They let people know of the existence of a particular text as well as pointing out its significance and importance. They also warn prospective readers about errors and deficiencies (Calef 1964). So many new publications are appearing that we need to be selective about what we read. Just as the recommendation of a trusted friend about a movie helps you decide whether to see a specific film or not, so too a carefully written review by a critical colleague helps academics and business people choose texts to read.

What are Your Review Markers Looking For?

People who read reviews, including those marking your review, typically want *honest* and *fair* comments on:

- what the reviewed item is about **(Description)**
- details of its strengths and weaknesses **(Analysis)**
- its contribution to the discipline **(Evaluation)**

The next few pages of guidelines and advice will help you to deal with these issues. The advice is structured around a series of quite specific questions most review readers want answered and which form the basis of the assessment schedule at the end of this chapter. These expectations are likely to be shared by most people commenting on a review you write for assessment. In beginning it is also worth noting that a review should itself be interesting as well as informative. So, while your assessor will probably expect the following issues to be addressed in the course of your review, there are no rules concerning the order in which they should be presented. Instead, the material should be set out in a manner which is both comprehensive and interesting. You are advised to read a few reviews in professional journals to see how they have been laid out and ordered before you write your own. Some good examples include Kearns (1994), Smith (1995), and Gulley (1995).

Description: what is the reviewed item about?

Description is an important part of a review. You should imagine that your audience has not read the text you are discussing and that their only knowledge of it will come from your review. Give them a comprehensive but concise outline of the text's content and character. However, do not make the mistake of devoting almost all of the review to description—if you do that, your reader might as well go to the original text! As a rule of thumb, try to keep the description to less than half the total length of the review.

Full bibliographic details of the text provided

You should provide a full and correctly set out reference to the work under review so that others may consult or purchase it (and so your marker knows that you have reviewed the correct text). Readers will be interested to know who the authors/editors are, the name of the publishing house, and where the volume was published. It is sometimes helpful to state how many pages are in the text (and if a book, its purchase cost, although this is rarely required for classroom reviews). An example of a complete book review reference is:

Lawrence, G., Vanclay, F. & Furze, B. (eds), 1992, *Agriculture, Environment and Society: Contemporary Issues for Australia*, MacMillan, Melbourne, pp. xiii and 337. A$32.95 (ppr), A$64.95 (bnd).

The text referred to in this example has 337 pages plus an additional 13 pages of introductory material (e.g. acknowledgements, title page, content pages, notes on contributors). The paperback edition costs $32.95 while the hardback is $64.95.

Bibliographic details are normally placed at the top of the review. The book review section of almost any academic journal will provide an example.

Sufficient details of author's background

If you consider it appropriate, and know of the author's expertise in the area they are writing about, provide a brief overview of their background and reputation (Northey & Knight 1992, p. 60). Have they written many other books and articles in this area and do they have an abundance of practical experience? Review readers unfamiliar with the subject area will often appreciate some information on an author's apparent credibility.

Text's subject matter identified clearly

Provide a short statement that lets your readers know what the reviewed text is about. For instance, is it a book on 'environmental problems in Nepal' or is it on 'development initiatives in rural Australia'?

In answering the question 'what is the text about?', you will probably find it appropriate to deal with another issue your reader/assessor expects you to address: 'what is the purpose of the text?' Distinguishing between the two matters can be a little difficult. The difference is clarified in the section below.

Purpose of the text clearly identified

Having stated what the text is about, you need to let the reader know precisely what the *aim* of the text is. The distinction between a text's subject matter and its purpose is illustrated by the following introductory sentences of Bourne and Ley's edited collection on the social geography of Canadian cities.

> This is a *book* about the places, the people and the practices that together comprise the social geography of Canadian cities. Its *purpose* is both to describe and to interpret something of the increasingly complex social characteristics of these cities and the diversity of living environments and lived experiences that they provide. (Ley & Bourne 1993, p. 3. Emphasis added)

You must think carefully about the distinction between a text's subject matter and its purpose. Usually a book or article will discuss some topic or example in order to make or illustrate a particular point. Thus, for example, two separate books about topics as different as the geography of medical malpractice and the geography of acid rain may actually have the same purpose: demonstrating that apolitical activities in one country may have profound political implications elsewhere.

Author's conceptual framework identified correctly

Texts are written from a particular perspective. Each author has a way of viewing the world and of arranging their observations into some specific and supposedly comprehensible whole. That means of thinking about the world is known as a conceptual framework. You might imagine a conceptual framework to be rather like the text's skeleton upon which the flesh of words and evidence is supported. As a reviewer, one of your tasks is to expose that skeleton, letting your reader know how the author has conceived the issues being addressed. How have they made sense of that part of the world they are discussing? In your review, you might combine identification and a critique of the conceptual framework. Does the author make inappropriate assumptions and are there inconsistencies, flaws, and weaknesses in the intellectual skeleton?

Succinct review of the text's content provided

Readers want some idea of what is in the book/article. This might require stating what is in various parts of the book and how much space is devoted to each section. You might want to integrate the summary of content with your evaluation of the text or you may wish to keep it separate (Northey & Knight 1992, p. 61).

Intended readers identified accurately

The reader of a review is usually interested to know what sort of audience the author of the original text was addressing. In many cases the author will include a statement on the intended readership somewhere early in the volume. For example, in the Preface to *The Slow Plague*, Peter Gould (1993, pp. xiii-xiv) says that his book is

> ...one of a series labeled (sic) *liber geographicus pro bono publico*—a geographical book for the public good, which sounds just a bit pretentious until we translate it more loosely as 'a book for the busy but still curious public.'

Identifying the intended audience serves at least two important purposes. First, you will be helping your readers decide whether the original text will be of any relevance to them. Second, you are providing yourself with an important foundation for writing your critique of the text (see the notes in the next section on style and tone of writing). For example, from Gould's statement above, it is reasonable to conclude that his book ought to be easily read, stimulating, and written for a lay audience. If it is not, there is an important flaw in the book. A review of a book should be written with the relationship between intended audience and content/style in mind.

Analysis: details of strengths and weaknesses

So far you have given your reader-assessor a few basic descriptive details about the text you are reviewing. Now you need to let everyone know what you consider to be the weaknesses *and* strengths of the text. You do not need to be entirely negative about a work you have read to show how much you know. If you believe that the material you are reviewing has no significant weaknesses, you should say so. However you should also point out its specific strengths.

As Calef (1964) observes, analysis is usually the weakest feature of book reviews.

> All authors deserve sympathetic, appreciative analyses of their books; too few authors get them. Many reviewers concentrate on the authors' mistakes and discuss the books as they should have been written.

It is critical that your review be written with consideration given to the aims of the original text's author. *Analyse the book on the author's terms*. Has the author achieved their aims?

In your analysis you should above all 'be fair, be explicit, be honest' (Calef 1964). To these ends you should explain why you agree or disagree with the author's methods, analysis, or conclusions.

In organising their analysis of a text, many reviewers point out its utility and successes first and then move on to point out the deficiencies. You may find that pattern a useful one to follow, but it is by no means a presentation prescription.

Text's contribution to your understanding of the world/discipline identified clearly

Begin with the assumption that the text's author has some useful contribution to make, rather than whether you agree or disagree with that contribution. What is that contribution? Has the author helped *you* to make sense of things? What has been illuminated? For example, a book on 'landscapes as texts' you are reviewing might have made you think differently about the ways in which you look at cities and towns around you. You should let the readers of your review know this.

Clear statement on achievement of text's aims

Think carefully about the author's stated or implicit objectives and compare those with the content of the text. Do they match one another? You would be derelict in your duty as a reviewer if you had stated what the reviewed text's aims were but failed to say whether or not they had been met. (Imagine how frustrated you would feel if someone told you there was an article in a magazine about how to make a lifetime fortune in 21 days and when you read those pages you found that the article failed to deliver. Back to Austudy, Ma and Pa or selling the car!)

Text's academic/professional functions identified clearly

What educational, research or professional functions might the text fulfil? For example, is the book or article likely to be a useful resource for people in the same class as you, for other undergraduate students, or for leaders in the field?

Accentuate the positive. For example, an author may think that their book's audience ought to be final year undergraduate students, but you as an undergraduate student reviewer believe the book would better serve a first year audience. Rather than simply stating that the text is inappropriate for its intended audience, let your reader know which audience the text might best serve. Make clear what can be obtained by using or reading the book and what it can best be used for.

Text's organisation commented on fairly

Say whether you believe the reviewed text is well organised or not. Within this, think about the ways in which the book/paper is subdivided into chapters/sections. Do the subdivisions advance the book's purpose or are they obstructive? Do they break up and upset an intellectual trajectory? Of course, you must also support your standpoint. For instance, you might criticise a book on urban settlement history in New Zealand for providing all of its examples before it offered a conceptual basis for making sense of those case studies. For most readers, the case studies would be somewhat meaningless without some prior idea of the way in which the author interpreted them.

Text's evidence evaluated critically

An author's evidence should be reliable, up-to-date, drawn from reputable sources, and should support claims made in the text. Would the results of the original text stand up to replication (i.e. if the study was done again, is it likely that the results would be the same?) or corroboration (i.e. are the results of the study substantiated by other related evidence?).

Make clear your assessment of evidence used in the text. Is it compelling or not? If, for example, you are sceptical about the repeated use of quotations drawn from the *National Enquirer* to support a case about bribery and gerrymandering in Queensland, you should let the reader know. Back up your assessment of the evidence with reasons for your conclusion. If you are able to, and if it is necessary, suggest alternative and better sources of evidence.

Text's references evaluated critically

A commentary on the comprehensiveness of references is given in relation to matters of evidence discussed above. Readers and assessors of your review will wish to know if the reviewed work has covered the available and relevant literature reasonably well. If there are major shortcomings in the references, it is possible that the text's author may not be fully aware of material that might have illuminated their work. If you are new to geography and environmental science you might protest, sometimes quite justifiably, that you cannot offer a meaningful judgment about the strength of the reference material. Nevertheless, you ought to be thinking about this question and answering it where possible.

Style and tone of writing in the text evaluated critically

Among other things, readers of a review may be trying to work out whether to buy or read the reviewed text. Therefore, questions foremost in the mind of some will be: 'is the text written clearly?' and 'is it interesting to read?' (Northey & Knight 1992, p. 61). Is the text's writing repetitious? Detailed? Not detailed enough? Is the style clear? Is it...plodding, jargon-laden, flippant? (Northey & Knight 1992, p. 62). If you think there are problems with the style of writing, it is appropriate to support your criticisms with a few examples.

You should also comment on the tone of the book. Let your reader/assessor know if the text is only accessible to experts in the field or if it is a coffee-table publication. Of course, criticisms about the tone of the text should be written with reference to the intended audience. It would probably be unfair, for instance, to condemn a book written for an audience of experts on the grounds that it used technical language.

Quality of supplementary material (e.g. tables, figures, plates) reviewed competently

Many books and articles in geography and environmental science make extensive use of supplementary material such as tables, plates and figures. In reviewing the text, you might comment on the quality of these materials and their contribution to the text's message. Graphic and tabular material in the text being reviewed should be relevant, concise, large enough to read, comprehensible and the details of sources should be provided.

Other deficiencies/strengths in the text identified correctly and fairly

After a close reading of the text, you should be able to identify any weaknesses and strengths you have not already discussed. Remember: you do not *have* to find things wrong with the text you are reviewing. Being critical does not require you to be negative. Similarly, you should not pick out minor problems within the text and suggest that they destroy the entire book. However, if there are genuine problems be explicit about what they are, providing examples where possible.

Evaluation: contribution to the discipline

In the first section of the review you described the text being reviewed. You then went on to outline the text's strengths and flaws. Now you have to make a judgment. Is the text any good? You may find it useful to be guided by the central question: *would you advise people to read (or buy) the text you have reviewed?*

Text compared usefully with others in the field

If you have the necessary expertise in the field, appraise the text being reviewed in terms of alternatives to its use or to work already available. To facilitate this, it may be helpful to consult library reference material to see, for example, how many other volumes on the same or a similar topic have been produced recently. Compare the book/article factually with its predecessors. What subjects does it treat that earlier volumes did not? What does it leave out? (Northey & Knight 1992, p. 60). Remember to cite correctly any additional sources you use.

Valid recommendation on the worth of reading the text provided

A fundamental reason for writing reviews is to let readers know whether a particular book or article is worth consulting. Your recommendation should be consistent with the preceding analysis of its strengths and weaknesses. For example, it would be inappropriate to criticise a book or article mercilessly and then conclude by saying that it is an important contribution to the discipline and should be read by all geographers and environmental scientists. It is worth restating a point made earlier. In your review *be fair, be explicit, be honest.*

Written expression and presentation of the review

Various sections of the review of appropriate length

One additional point assessors will consider in marking a review is the balance between description, analysis and evaluation. Many novice reviewers make the mistake of devoting almost their entire review to description. As was pointed out before, unless there are special reasons, the description should not usually be more than half of the total length of the review.

Your review must be understood by its intended audience. For that reason, you can expect an assessor to comment critically on the quality of your written expression. Detailed advice on the written expression assessment criteria shown in the assessment schedule for reviews, but not outlined here, can be found in the chapter on Essay Writing.

Writing a Summary or Precis

What is a summary or precis?

A summary or precis is simply a brief outline of the content and argument of a book/article/chapter. Lecturers will usually ask you to write a summary to ensure that you have read and understood an important article or book. Note, therefore, that the task of writing a summary is a task of comprehension, not recitation.

What is the purpose of a summary or precis?

In writing a precis you simply have to record, as accurately as possible, and within any prescribed word limit, your understanding of what the author has written (Northey & Knight 1992, p. 58). Unlike a review, a summary does not contain interpretation of the issues raised. It is not evaluative. There is no need for you to provide your reaction to the ideas of the author.

Your precis must accurately *re-present* the text in abridged form. Imagine you are trying to write a scaled-down version of the original. Brevity and clarity are critical ingredients of this type of work. Let your reader–assessor know in as few words as possible what the summarised text is about. Do not prepare a summary which is as long as the article itself. Imagine your audience sitting opposite you with a bored glaze about to appear across their eyes. Spare them the details. Give them enough information to understand what the text is about but not so much that the central theme is lost.

This leads us to one of the most common shortcomings of summaries: failing to state the original author's *main argument*. Imagine someone has asked you to tell them about a movie you have recently seen. One of the most important things they will want to know, and in relatively few words, is *essential details of the plot*. For example, readers probably do not want to know the names of Cinderella's evil sisters, the colour of her ball-gown, and the temperament of the horses. Instead, they want to know the essence of the story which brought Cinderella together with her Prince Charming.

Once you have written your precis, re-read it with the following question in mind: 'could I read this precis aloud to the author in the honest belief that it accurately summarises their work?' If the answer is no, modify your precis.

What is the reader of a summary/precis looking for?

Full bibliographic details of the text provided

At the start of your summary, provide the reader with a full reference to the work being summarised. See the notes for Writing a Review for details.

> Text's subject matter identified clearly

See the notes for Writing a Review for details.

> Purpose of the text clearly identified

See the notes for Writing a Review for details.

> Relative emphases in the precis match emphases in original text

In the summary you should give the same relative emphasis to each area as did the original text's author (Northey & Knight 1992, p. 59). If, for example, two-thirds of a paper on monitoring water pollution from New South Wales ski fields was devoted to a discussion of the legalities of obtaining the water samples, your summary should also devote two-thirds of its attention to that issue. This helps to provide your reader with an accurate view of the original text.

> Order of presentation in precis matches that of original text

Just as you should give the same emphasis to each section as the original text's author, so you should follow the article/book's order of presentation and its chain of argument (Northey & Knight 1992, p. 59; South Australian College Advanced Education 1989). Make sure you have presented enough material for a reader to be able to follow the logic of each important argument. You will not be able to provide every detail. Present only the critical connections.

> Key evidence supporting the original author's claims outlined fully

Briefly mention the critical evidence provided by the author to support their arguments (Northey & Knight 1992, p. 59). There is no need to recount all of the data/evidence offered by the author. Instead, refer to the material which was the most compelling and convincing.

Summary written in own words

Write the summary in your own words, although you may, of course, elucidate some points with quotes from the original source. Do not construct a precis as a collection of direct quotes from the text you are summarising.

Preparing an Annotated Bibliography

What is an annotated bibliography?

An annotated bibliography is a list of reference materials such as books and articles in which author, title and publication details are provided (as in a bibliography) and in which each item is summarised. Annotated bibliographies are customarily set out in alphabetical order (by authors' surnames) and usually include a short synopsis of the text followed by a concise critique (i.e. your opinion of the work). Some bibliographies are, however, written in the form of a short essay.

What is the purpose of an annotated bibliography?

Annotated bibliographies are used to provide people in a particular field of inquiry with some commentary on (new) books and articles available in that field. Alternatively, annotated bibliographies may be used to provide a newcomer with an insightful review of reference material available.

What is the reader of an annotated bibliography looking for?

Although the content may vary depending on the purpose of the list, readers of annotated bibliographies *typically* expect to see three sets of information:

- **Details** Full bibliographic details (author, date of publication, title, volume or edition number, pages, place of publication etc.).
- **Summary** A clear indication of the content (and argument) of the piece
- **Critique** Critical comment on the merits and weaknesses of the publication or of its contribution to the field of study.

Some examples of entries in annotated bibliographies

1 Beasley, V. 1984, *Eureka! or How to be a Successful Student*, Flinders University of South Australia.
This book contains strategies to help students become more effective learners and covers such areas as elements of learning, establishing a purpose, surveying, questioning, collecting, integrating, reviewing, communicating, effective reading, and writing essays. This is an easy-to-read reference that provides positive and extremely well researched strategies for improving the quality of learning. (South Australian College of Advanced Education 1989)

2 Moss, P. 1993, 'Focus: feminism as method', *Canadian Geographer*, vol. 37, no. 1, pp. 48–49.
This short introduction to a series of papers makes some contentious claims about 'feminist' and 'masculine' methods in geography. In general, and despite some of her claims to the contrary, Moss appears to equate quantitative/positivist geography with masculinism; qualitative geography with feminism. Not only is this misleading, but it also rests on sexist conceptions of femininity and masculinity. (Hay 1995a)

3 I know of no geographical research on the condom, and virtually nothing on the geography of sexual relations, although the fine and pioneering work by a geographer:
Symanski, R. 1981, *The Immoral Landscape: Female Prostitution in Western Societies*, Butterworths, Toronto;
is increasingly referenced by other human scientists. Numerous short reports on condom use and propagation are given in almost all issues of *WorldAIDS*, while many articles and reports in the 'AIDS Monitor' of *New Scientist* deal with condom use. (Gould 1993, p. 215)

Book Review Assessment Schedule

Student Name: Grade: Assessed by:

The following is an itemised rating scale of various aspects of book/article review performance. Sections left blank are not relevant to the attached assignment. Some aspects are more important than others, so there is no formula connecting the scatter of ticks with the final grade for the assignment. Ticks in either of the two boxes left of centre means that the statement is true to a greater (outer left) or lesser (inner left) extent. The same principle applied to the right-hand boxes means that the topic was treated somewhat superficially in the assignment. If you have any questions about the individual scales, comment, final grade or other aspects of this assignment, please see the assessor indicated above.

Aspect 1 Description

Full bibliographic details of the text provided					Insufficient bibliographic details
Sufficient details of author's background					No details of author's background
Text's subject matter identified clearly					Text's subject matter poorly or inadequately identified
Purpose of the text clearly identified					Text's purpose unstated or unclear
Author's conceptual framework identified correctly					Little or no attempt to identify conceptual framework
Succinct review of the text's content provided					Over-long/inadequate review of content
Intended readers identified accurately					Readership not identified

Aspect 2 Analysis

Text's contribution to understanding of the world/discipline identified clearly					Little or no reference to text's contribution
Clear statement on achievement of text's aims					Text's aims not identified or identified incorrectly
Text's academic/professional functions identified clearly					Text's functions not identified or identified incorrectly
Text's organisation commented on fairly					Little or no comment on organisation
Text's evidence evaluated critically					Little or no comment on evidence
Text's references evaluated critically					Little or no comment on references
Style and tone of writing in the text evaluated critically					Little or no comment on style and tone
Quality of supplementary material (e.g. tables, maps, plates) reviewed competently					Little or no comment on supplementary material
Other deficiencies/strengths in the text identified correctly and fairly					Other evident weaknesses/strengths not identified

Aspect 3 Evaluation

Text compared usefully with others in the field					Little or no effort to contrast text with other comparable texts in the field
Valid recommendation on the worth of reading the text provided					No recommendation provided or recommendation inconsistent with earlier comments

Aspect 4
Written Expression, References and Presentation of the Review

Various sections of the review of appropriate length					Major imbalances evident
Fluent piece of writing					Clumsily written
Succinct writing					Verbose and/or repetitive
Grammatical sentences					Many ungrammatical sentences
Correct punctuation					Much incorrect punctuation
Correct spelling throughout					Much incorrect spelling
Legible, well set-out work					Untidy and difficult to read
Reasonable length					Over/under length
Correct and consistent in-text referencing style					Incorrect and/or inconsistent referencing style
Reference list correctly presented					Errors and inconsistencies in reference list

Assessor's comments

Summary/Precis Assessment Schedule

Student Name: Grade: Assessed by:

The following is an itemised rating scale of various aspects of summary / precis performance. Sections left blank are not relevant to the attached assignment. Some aspects are more important than others, so there is no formula connecting the scatter of ticks with the final grade for the assignment. Ticks in either of the two boxes left of centre means that the statement is true to a greater (outer left) or lesser (inner left) extent. The same principle applied to the right-hand boxes means that the topic was treated somewhat superficially in the assignment. If you have any questions about the individual scales, comment, final grade or other aspects of this assignment, please see the assessor indicated above.

Full bibliographic details of the text provided					Insufficient bibliographic details
Text's subject matter identified clearly					Text's subject matter poorly or inadequately defined
Purpose of the text clearly identified					Text's purpose unstated or unclear
Relative emphases in the precis match emphases in original text					Little or no correspondence in text-precis emphases
Order of presentation in precis matches that of original text					Little or no correspondence in text-precis order of presentation
Key evidence supporting the original author's claims outlined fully					Little or no reference to summarised text's evidence
Summary written in own words					Precis constructed largely from quotes from summarised text

Written Expression and Presentation of the Precis/Summary

Fluent piece of writing					Clumsily written
Succinct writing					Verbose and/or repetitive
Grammatical sentences					Many ungrammatical sentences
Correct punctuation					Much incorrect punctuation
Correct spelling throughout					Much incorrect spelling
Legible, well set-out work					Untidy and difficult to read
Reasonable length					Over/under length
Correct and consistent in-text referencing style					Incorrect and/or inconsistent referencing style
Reference list correctly presented					Errors and inconsistencies in reference list

4

Preparing maps, figures, and tables

As Bertrand Russell once said, most of us can recognise a sparrow, but we'd be hard put to describe its characteristics clearly enough for someone else to recognise one. Far more sensible to show them a picture.

(Rowntree 1990, p. 193)

Geographers would be lost without maps. This chapter discusses the importance of maps and other graphic devices (as well as tables) and offers advice on the ways in which they are produced.

Why Communicate Graphically?

In geography and environmental studies, words or numbers alone are often not sufficient to communicate information effectively. Graphic communication allows a large amount of information to be displayed succinctly and absorbed readily. Effective illustrations can help a reader achieve a rapid understanding of an argument or issue.

Graphs employ human powers of visual perception and pattern recognition, which are much better developed than our capacity to uncover meaningful relations in numerical lists (Krohn 1991, p. 188). Often we see things in graphic form which are not apparent in tables and text. Krohn argues that graphs reveal relationships that allow both the numbers upon which they are based and the concepts by which we understand those numbers to be reinterpreted. As such, graphs are critical interactive sites for comprehending the world around us.

In addition to their intellectual functions, graphics can enhance various forms of technical writing in different ways (Eisenberg 1992, p. 81):

- in reports: graphs and tables summarise quantitative information, freeing up text for comments on important features
- in instructions: graphics may help people to understand principles behind the operation of some process or the characteristics of a phenomenon
- in oral presentations: charts, tables and figures relieve monotony, help guide the speaker and aid the audience's understanding of data.

Several different types of illustrative material can be used in essays, reports, posters, and seminars. This chapter discusses the character and construction of different types of graphic. It is important first to introduce a few general guidelines.

General Guidelines for Clear Graphic Communication

The principal aim of data graphics is to display data accurately and clearly (Wainer 1984, p. 137). Good graphics should be:

1 Concise

Graphics should present only that information which is relevant to your work and required to make your point. Critically review the data you are going to portray to find out what they 'say' and then let them say it graphically with the minimum of embellishment (Wainer 1984, p. 147). If you reproduce an illustration or table from published sources or other research you may need to redraw or rewrite it to remove irrelevant details.

2 Comprehensible

Your audience must understand what the graphic is about. Central to this end is a good title and effective labelling. An illustration must have a clear and complete title which answers 'what', 'where', and 'when' questions. Effective data labels and axis labels are:

- legible and easy to find;
- easily associated with the axis/object depicted (they should be close together);

- readable from a single viewpoint. A reader looking at the graphic should be able to read the text without having to turn the page sideways. On maps, the general point about viewpoint should be heeded but labels should also be aligned along linear features (e.g. roads), extend over areal features (e.g. seas, suburbs) and should consistently locate point features (e.g. all point features labels at the four o'clock position relative to the item being labelled) (Gerber 1990–1, p. 28).

While a graphic should be fully labelled, care should be taken to ensure that data regions are as clear as is practicable of notes, axis markers, and keys. In short, the graphed information should be clear and easy to read. If the graph displays two or more data sets they must be easily distinguished from one another. The Cartography Specialty Group of the Association of American Geographers (1995, p. 5) suggests that graphs include no more than *four* simultaneous symbols, values, or lines and that each line or symbol be sufficiently different from the others to facilitate easy discrimination.

Comprehensibility of a graphic is also fostered through effective use of the data region (i.e. that part of the graph within which the data is displayed). Choose a range of axis scale marks that will allow the full range of data being depicted to be included, but also ensure that the scale allows the data to fill up as much of the data region as possible. If you take photographs, this principle will be familiar to you. Just as good photos will usually 'fill the frame', so a good graph will typically fill the data region.

3 Independent

Graphics should stand alone. Someone who has not read the document associated with the graphic should be able to look at the table or illustration and understand what it means. Graphics should also be independent of one another.

4 Referenced

Sources must be acknowledged. Use the conventions of an accepted referencing system to refer fully to sources of data and to graphics. A reference list at the end of your work should provide the full bibliographic details of data and graphic sources. Each graphic should be accompanied by summary bibliographic details (author, date, page in the case of the author-date system) or a note identifier allowing the reader to find out where the graph or the data upon which it is based came from.

Ensure that the reference is to the source you used, and not that of the author of the text you are borrowing from. For example, imagine you are copying a penguin population graph you found in a 1996 book by Dr. Emperor. Emperor had, in turn, cited the source of her graphed data as the Argentinian Penguin Research Foundation. Following the author-date system, the graph you present in your work would be referenced as (Argentinian Penguin Research Foundation, in Emperor 1996, p. 12). The reference would not simply be to Emperor. Of course, there would also be a full reference to Emperor's work in the list at the end of your paper. See the chapter on referencing for further information.

Different Types of Graphic

The following sections of this chapter briefly describe various forms of graphic communication and provide some advice on their construction. Table 4.1 provides summary of the major forms of graphic discussed and their nature and function. If you have data or other information you wish to depict graphically you might find it useful to consult the table for some indication of the specific form of graphic device you might employ.

Table 4.1: Types of graphic and their nature/function

Type of Graphic	*Nature/Function*
Scattergram	Graph of point data plotted by (*x,y*) co-ordinates. Usually created to provide visual impression of direction and strength of relationship between variables.
Line Graph	Values of observed phenomena are connected by lines. Used to illustrate change over time.
Bar Chart	Observed values are depicted by one or more horizontal or vertical bars whose *length* is proportional to value(s) represented.
Histogram	Similar to bar graph, but commonly used to depict distribution of a continuous variable. Bar *area* is proportional to value represented. Thus, if class intervals depicted are of different sizes, the column areas will reflect this.
Population Pyramid	Form of histogram showing the number or percentage of people in different age groups of a population.
Pie (Circle) Chart	Circular shaped graph in which proportions of some total sum (the whole 'pie') are depicted as 'slices'. The area of each 'slice' is directly proportional to the size of the variable portrayed.
Logarithmic Graph (log-log and semi-log)	Form of graph using logarithmic graph paper. Key intervals on logarithmic axes are exponents of ten. Log graphs allow depiction of wide data ranges.

Table	Systematically arranged list of facts or numbers, usually set out in rows and columns. Presents summary data or information in orderly, unified fashion.
Dot Map	Uses dots to illustrate spatial distribution of discrete data by unit of occurrence (e.g. one dot represents 1 person) or some multiple of those units (e.g. 1 dot represents 1,000 sheep).
Choropleth Map	Shaded or crosshatched map used to display statistical distributions (e.g. rates, frequencies, ratios) on the basis of areal units such as nations, states, and regions.
Isoline Map	Shows sets of lines connecting points of known, or estimated, equal values (e.g. elevation, barometric pressure, temperature).

Scattergrams

A scattergram is a graph upon which one plots point data (see figure 4.1 for an example[1]). This is commonly done to gain some visual impression of the direction and strength of a relationship between variables.

Usually one set of data is a *dependent* variable quantity. This is depicted on the vertical *y-axis* and the other set of data is an *independent* range of points, depicted on the horizontal *x-axis* (Windschuttle & Windschuttle 1988, pp. 274–275). To illustrate the difference between independent and dependent variables, consider the relationship between precipitation levels and costs associated with flooding. We can assume that damage costs associated with flooding will usually depend on the amount of rainfall. Thus, rainfall is the independent variable (x-axis) and damage costs are dependent (y-axis). Or, the severity of injuries associated with a motor vehicle accident (dependent variable, y-axis) tends to increase with motor vehicle speed (independent variable, x-axis).

After points are located on the scattergram, you might draw a 'line of best fit' through the points by eye (i.e. your visual impression of the relationship expressed in the form of a line through the points). This line may be calculated mathematically and the regression equation expressed on the graph (see figure 4.1).

1 In a deliberate strategy, all graphs in this book have been drawn using Microsoft Excel. While other, more powerful software packages for producing graphics exist, Excel is commonly available and produces adequate figures for most undergraduate assignments. It is readily available in most universities and students familiar with computers should be able to produce graphics comparable with (or better than) any of those shown in this chapter. Hand-drawn figures can easily be drawn to the same standard.

Figure 4.1: Example of a scattergram. Male life expectancy and rates of natural increase, selected countries (1994)

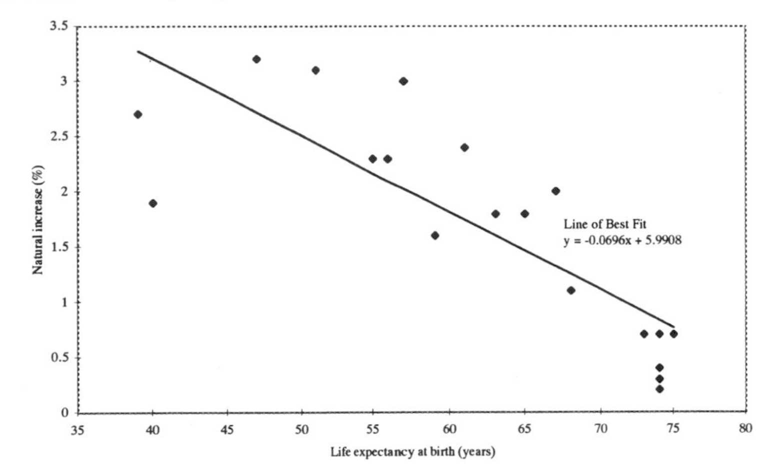

Data Source: World Almanac (1995)

Figure 4.2: Example of a line graph. Number of items on South Australian Heritage Register, 1980–1992.

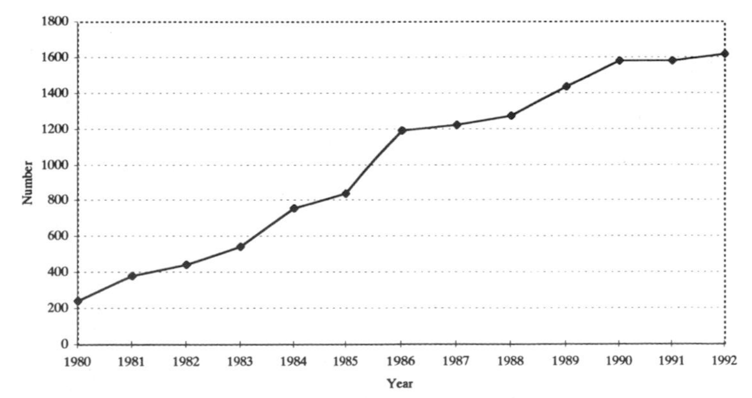

Data Source: Department of Environment and Land Management (1993, p.263)

Line graphs

Typically, line graphs are used to illustrate continuous changes in some phenomenon over time, with any trends being shown by the rise and fall of the line. Line graphs may also show the relationship between two sets of data. Figures 4.2 and 4.3 are examples of line graphs.

Figure 4.3: Example of a line graph. Permanent arrivals and departures, Western Australia (March 1993–June 1994).

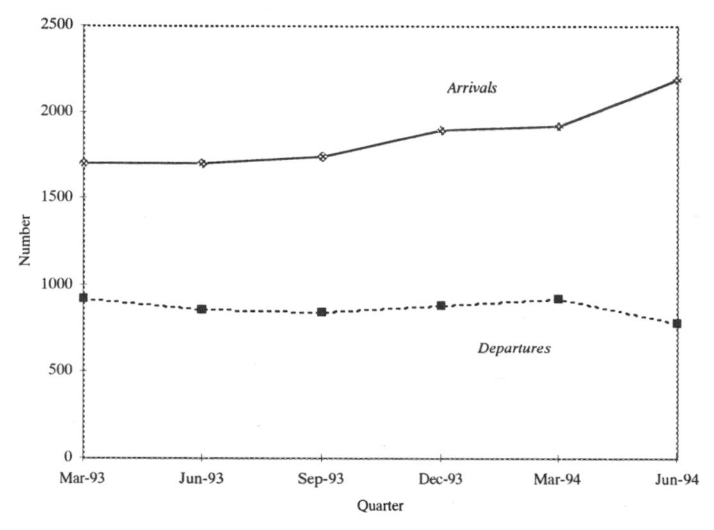

Data Source: ABS (1994e, p.4)

Do not use a line graph if you are dealing with disconnected data (Eisenberg 1992, p. 97). For example, if you have air pollution data for every second year since 1945, the information should be graphed using a bar chart because a line graph would incorrectly suggest that you have data for each intervening year.

Construction of a line graph

Plot each (x,y) data point for your data set(s). When all the data points are plotted, join the points associated with each data set to produce lines such as those shown in figures 4.2 and 4.3. The line you draw may be curved artistically or 'straightlined' (Mohan, McGregor & Strano 1992, p. 284). If a number of lines are depicted in one graph, ensure that they can easily be distinguished from one another (see figure 4.3) through use of colour, dotted lines, or labels.

Line graphs will sometimes compare things that have different measurements. This can be done by using vertical axes on the left and right sides of the graph to depict the different scales. Figure 4.4 provides an illustration of the use of multiple vertical axis labels.

Figure 4.4: Example of a graph using multiple vertical axis labels. Brisbane climate, 1994

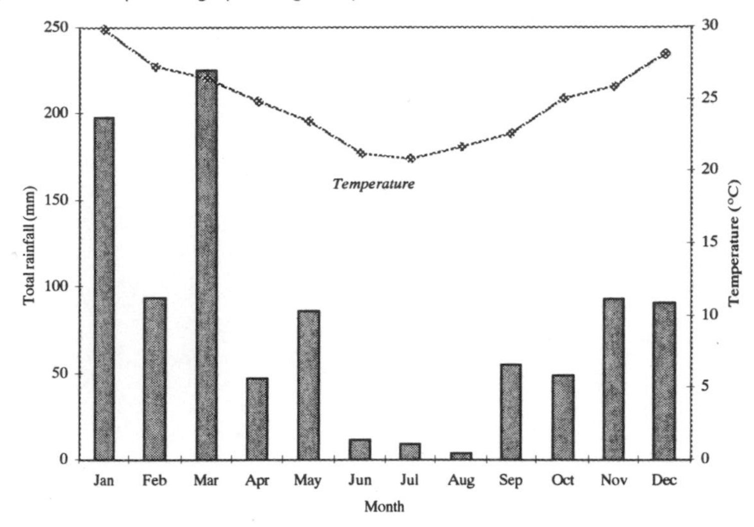

Data Source: ABS (1994d, p.22)

Figure 4.5: Example of a horizontal bar graph. Percentages of electricity mains underground, Australian states, 30 June 1991

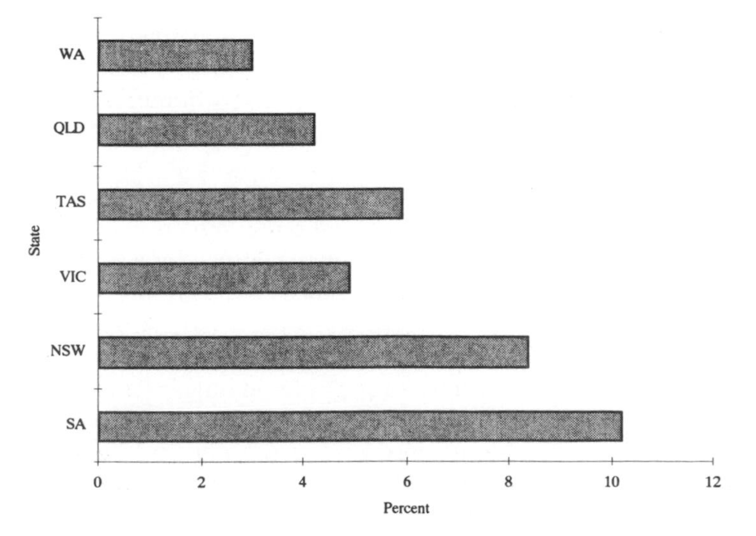

Data Source: Department of Environment and Land Management (1993, p. 215)

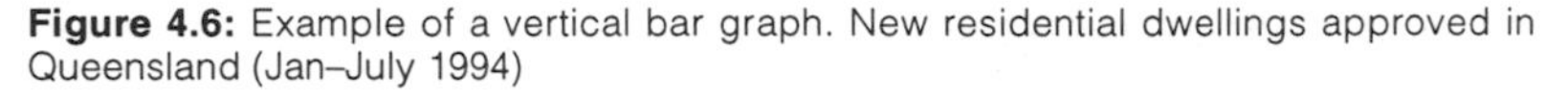

Figure 4.6: Example of a vertical bar graph. New residential dwellings approved in Queensland (Jan–July 1994)

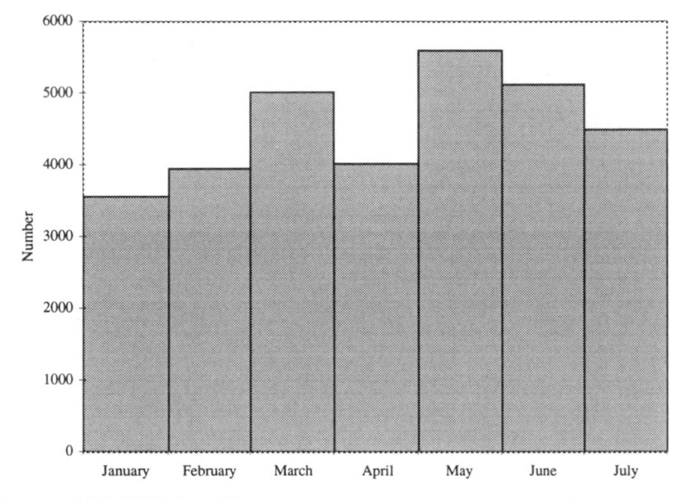

Data Source: ABS (1994d, p.13)

Bar Charts

Bar charts are of two main types: *horizontal* and *vertical* (ABS 1994g, p. 108). Figures 4.5 and 4.6 show each type respectively. Horizontal bar graphs usually represent a single period of time, whereas column graphs may represent similar items at different times (Moorhouse 1974, p. 67).

Although it sounds obvious, note that in a bar graph, the *length* of each bar is proportional to the value it represents (Coggins & Hefford 1973, p. 66). It is on that point that bar graphs differ from histograms, with which they are sometimes confused. Discussed later, histograms also use bars, but bars whose *areas* are proportional to the values depicted.

Bar charts are a commonly used and easily understood way of taking a snapshot of variables at one point in time, depicting data in groups, and showing the size of each group (Windschuttle & Windschuttle 1988, p. 278; Moorhouse 1974, p. 64). Figure 4.7 achieves all of these ends in a single graph.

Bar charts can also be used to show the components of data as well as data totals. See figure 4.8 for an example of such a *subdivided bar chart*. It is possible to go one step further and represent data in the form of a *subdivided 100% bar chart* (see figure 4.9 for an example).

Figure 4.7: Example of a bar chart. Methods of suicide by urban/rural location, Australia (1987–1991).

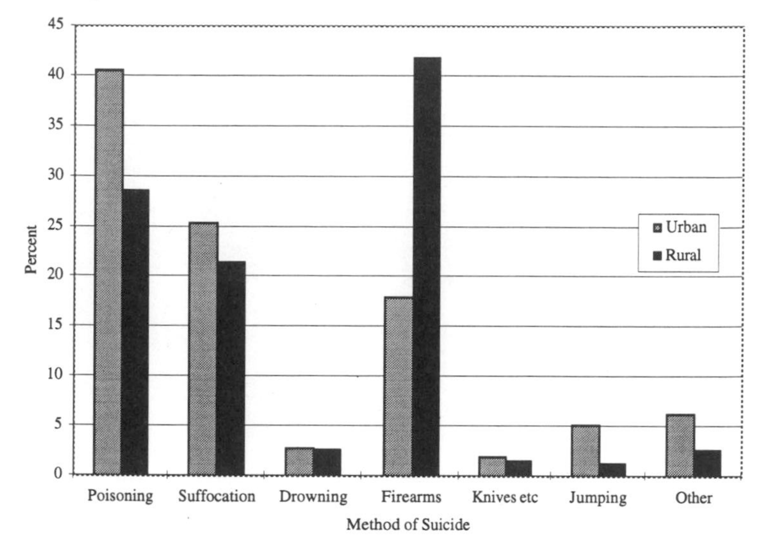

Data Source: ABS (1994b, p.59)

Figure 4.8: Example of a subdivided bar chart. Comparison of part-time and full-time employment in wholesale and retail trade, Australia (1973–1993).

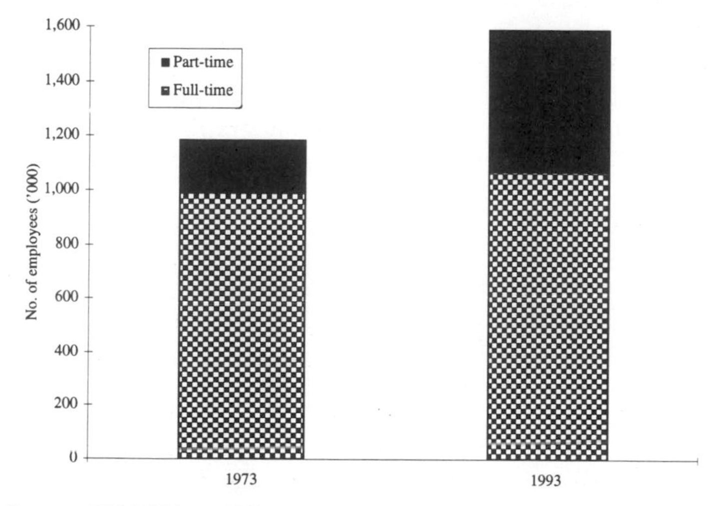

Data Source: ABS (1994a, p.105)

Figure 4.9: Example of a subdivided 100% bar chart. Paper and garden waste recycling by Australian households, May 1992.

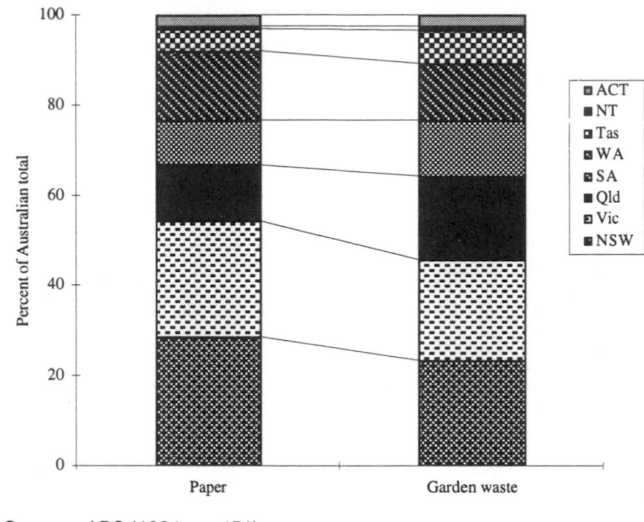

Data Source: ABS (1994a, p.451)

Figure 4.10: Example of a bar chart depicting positive and negative values. Average annual farm income, New South Wales, 1979–1986.

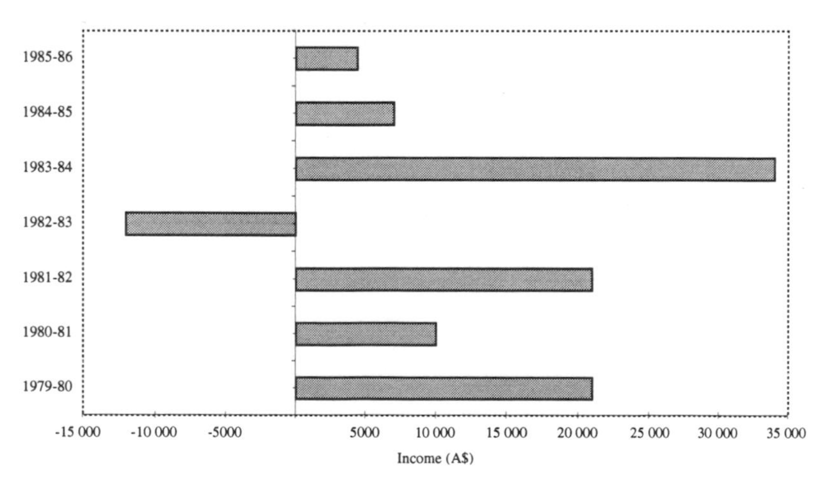

Source: Central Mapping Authority, New South Wales (1987, p.28)

These can be useful for depicting figures whose totals are so different that it would be almost impossible to chart them in absolute amounts (Moorhouse 1974, p. 66).

As Figure 4.10 illustrates, bar charts can be used to portray negative as well as positive quantities.

Construction of a bar chart

Examine the data which is to be graphed and select suitable scales for the graph's axes. In general, scales should commence at zero (Coggins & Hefford 1973, p. 66) although this is not critical. Label the axes.

When the chart is being designed, the sequence of the items to be depicted must be taken into consideration. In general, the items should be listed in order of importance to the viewer. However, in simple comparisons in a horizontal bar graph format it is best to arrange the bars in ascending order of length from bottom to top. Having said that, you must also be aware that some data sets are listed, by convention, in particular orders. For example, in Australia, Bureau of Statistics occupational groups are typically listed in the following order: Managers & Administrators, Professionals, Para-professionals, Tradespersons...Labourers & related workers, Inadequately described, Not stated. Similarly, industrial groups are usually listed in order through primary (e.g. farming), secondary (e.g. manufacturing), tertiary (e.g. retail) and quaternary (e.g. information transfer) divisions. Graphs should typically reflect such customary presentation forms. If you are not sure whether to set up a table in ascending order, speak to your lecturer.

The next step is to draw in the bars. Their width is a matter of choice, but should be constant within a graph. If you use different widths within the same graph, some readers may be led to believe that bar width, and hence area, is more important than length. Bars should be separated from one another, reflecting the discrete nature of the observed values (Jennings 1990, p. 18) and the space between the bars should be about one-half to three-quarters of their width. However, where the pattern of change is of greater importance than the individual values, no space between the observations is left at all (Coggins & Hefford 1973, p. 66). For example, compare figures 4.6 and 4.7. In figure 4.6 the pattern of change through the year is of more interest to the reader than is the specific value for each month. By contrast, the different, and therefore graphically differentiated, means of suicide portrayed in figure 4.7 are central to the chart's message.

Finally, add appropriate title, labels, key and reference.

Histograms

Histograms are mainly used to show the distribution of values of a continuous variable. A continuous variable is one which could have any conceivable value within an observed range (e.g. plant height, rainfall measurements, temperature) and may be contrasted therefore with discrete data in which no fractional numbers such as halves, quarters exist (e.g. plant and animal numbers). For examples of histograms, see figure 4.11. This figure shows three histograms drawn using the same data set but different class intervals. Class intervals are explained shortly.

Histograms may be confused with vertical bar charts or column graphs, but there is a technical difference. Strictly speaking, histograms depict frequency through the *area* of the column, whereas in a column graph frequency is measured by column *height* (ABS 1994g, p. 120). Thus, whilst histograms usually have bars of equal width, if the class intervals are of different sizes the columns should reflect this. For example, if one class interval on a graph was $0 to $9 and the second was $10 to $29, the second should be drawn twice as wide as the first.

The phenomenon whose size is being depicted is plotted on the horizontal x-axis. Frequency of occurrence is plotted on the vertical y-axis. The frequency is the number of occurrences of the measured variable within a specific class interval (e.g. number of hotels with rooms available in a given price category).

Construction of a histogram

As figure 4.11 illustrates, the method you choose to construct your histogram can have a significant effect on the appearance of the graph you finally produce. Figures 4.11(a), (b), and (c) were drawn using the same data (from table 4.2). However, each was drawn using different methods of calculating class intervals and frequency distributions. Figure 4.11(a) splits the data range evenly on the basis of the *number* of x-axis classes desired. Figure 4.11(b) shows the data on the basis of the desired *size* of the x-axis classes (in this case $55) and figure 4.11(c) is the product of *minimising in-class variations* while maximising between-class variations.

The first two methods of working out class intervals and frequency distributions which allow you to summarise the data to be depicted in your histogram require that you calculate the range of

Figure 4.11a: Example of histograms drawn using the same data but different class intervals.
(a) Hotel accommodation costs, Wellington (1994)

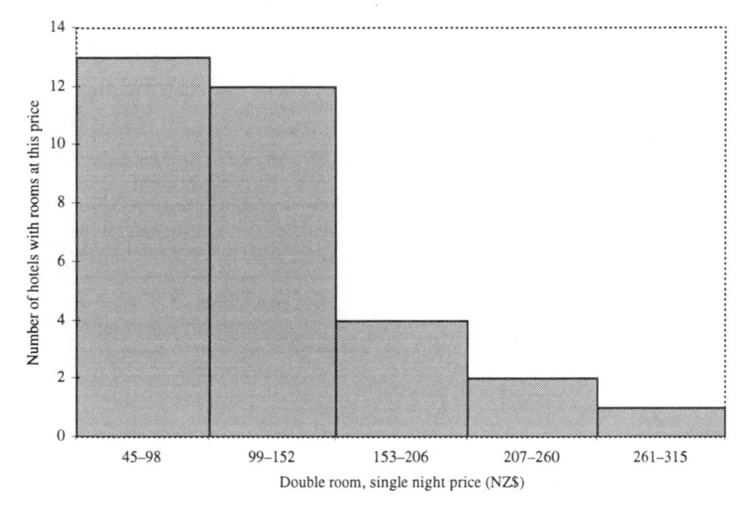

Data Source: New Zealand Tourism Board (1995)

(b) Hotel accommodation costs, Wellington (1994)

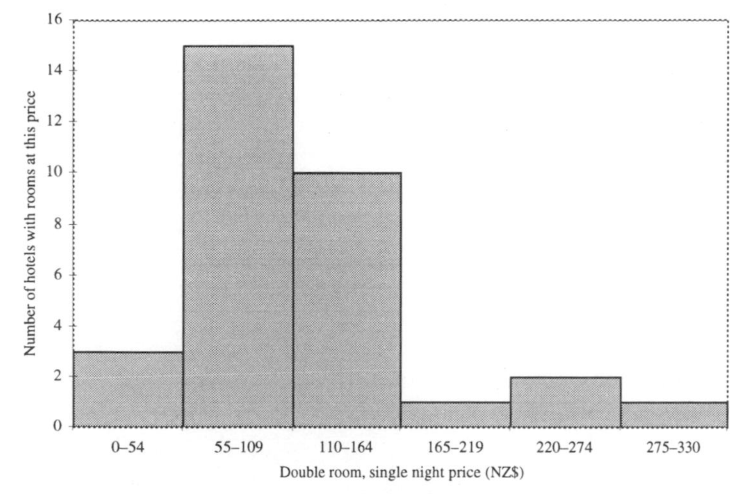

Data Source: New Zealand Tourism Board (1995)

(c) Hotel accommodation costs, Wellington (1994)

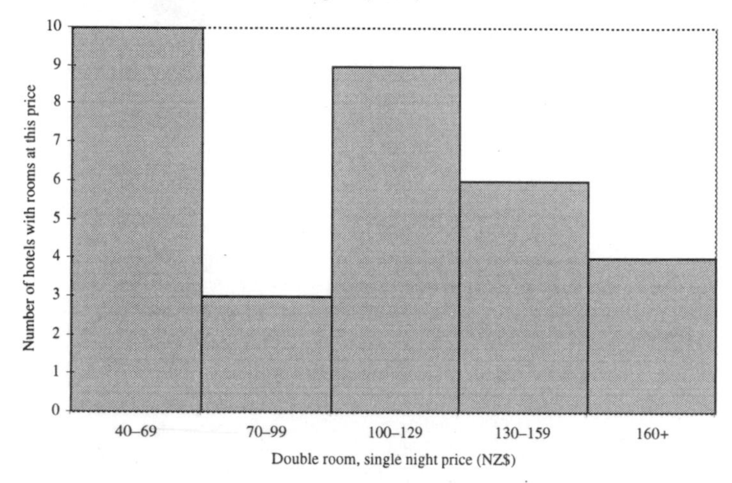

Data Source: New Zealand Tourism Board (1995)

the data set. Range is the difference between the highest data value and the lowest data value. To illustrate, consider the data shown in table 4.2 which displays the price of hotel accommodation in Wellington, New Zealand during 1994.

Table 4.2: (Data set for histogram construction) Single night, double room hotel accommodation rates (NZ$), Wellington 1994

109	253	118	56
112	124	60	45
95	162	90	59
45	198	100	50
136	156	315	50
156	105	144	65
253	118	101	55
105	152	50	80

Source: New Zealand Tourism Board (1995).

The most expensive room rate in Wellington in 1994 was $315, the lowest was $45. Therefore, the range is:

$315–$45 = $270

The next step is to calculate class intervals.

Methods of calculating class intervals

1 One common strategy for calculating class intervals is simply to divide the range by the number of classes you wish to portray. The

result will be a number of even size classes. For example, if we use the data in table 4.2, the range is $315–$45 = $270. You might have decided that you wish to have a histogram with five classes. Divide $270 by 5 and the result is an interval of $54. Thus, we have intervals of:

Class 1	$45–$98
Class 2	$99–$152
Class 3	$153–$206
Class 4	$207–$260
Class 5	$261–$315

The lowest class begins with the lowest value ($45 in this example). To find the *lower limit* of the *next* class we add $54, which produces a figure of $99. We then add $54 to $99 to produce the lower limit of the next class, $153, and so on. The *upper limit* of each class is found by subtracting 1 unit of the measurement form being used (e.g. $1, 1 cm, 1 m, 0.01 gram, 1 tonne) from the lower limit of the class above. The upper limit of the lowest class in the example is therefore $98. Repeat this procedure until the intervals for all classes are calculated. Note that *discrete class intervals* are used (e.g. $45–$98, $99–$152) rather than $55–$99, $99–$153). In this way there is no confusion about the class within which any data point is placed (e.g. in which class would you put a $99 room charge?)

2 An alternative, but closely related, strategy is the one followed in producing figure 4.11. Calculate the data range and then think about the character of the data set to be portrayed. Would it be useful to your audience to read the data on a graph which uses intervals of, for example, 10s or 50s rather than the 13s, 77s, $54s and other odd numbers which might be achieved by the simple division of the range by the number of desired class intervals as described in the preceding strategy? Similarly, would it be useful to commence or end the class intervals at some points other than those fixed by the high and low points of the data set? With these thoughts in mind, I chose the following intervals for the data in table 4.1.

Class 1	$0–$54
Class 2	$55–$109
Class 3	$110–$164
Class 4	$165–$219
Class 5	$220–$274
Class 6	$275–$330

The graph that resulted is figure 4.11. As you can see, class intervals that extended in value beyond the upper and lower limits of the data range were selected. The first class commences at $0 and the class interval is $55, which I thought useful when one is considering the matter of hotel accommodation costs in New Zealand.

3 Yet another technique of working out class intervals is to minimise in-class variations while simultaneously maximising between-class variations. Look for clusters of data points within the total data set and subdivide the data range using those natural breaks into equal size divisions which best discriminate between clusters. A useful tool in this process is the *linear plot*. Draw a horizontal line and affix to it a scale sufficient to embrace the maximum and minimum values of the data. Locate each of the data points on the scale with a short vertical line. If you are using the linear plot for presentation purposes, rather than for calculation only, you should also label each of the data points and provide a title and source. The plot will graphically portray the data distribution.

Figure 4.12: Example of a linear plot.

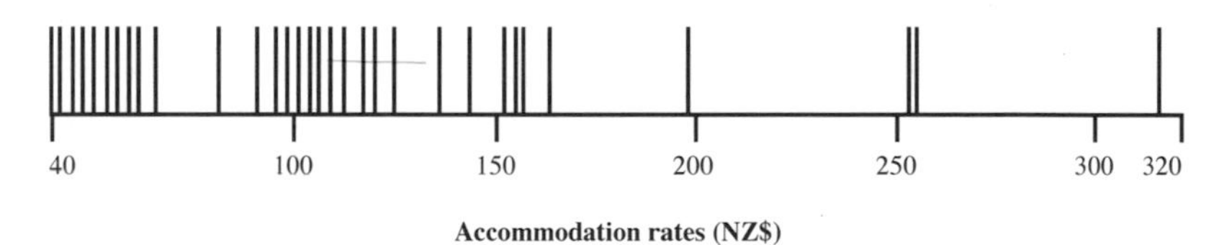

In this example, the data is clustered quite heavily in the range $50–$150. It might be appropriate to produce a histogram which breaks the data into the following ranges:

Class 1	$40–$69
Class 2	$70–$99
Class 3	$100–$129
Class 4	$130–$159
Class 5	$160

There is no definite rule governing the number of classes in a frequency distribution. Choose too few and information could be lost through a large summarising effect. That is, the picture will be too general. With a lot of classes, too many minor details may be retained, thereby obscuring major features.

Once you have worked out class intervals, the next step in the construction of a histogram is the creation of a frequency table which will allow you to work out the total number of individual items of data that will occur in a particular class.

Constructing a frequency table

As shown in table 4.3, constructing a frequency table is simply a matter of going through the set of data, placing a tally mark against the class into which each datum falls, then summing the tally to find the frequency with which values occur in each class. Table 4.3 uses the class intervals described on p.73.

Table 4.3: Frequency table of data from table 4.2

Classes ($)	*Tally of occurrence*	*Frequency*
0–54	III	3
55–109	~~IIII~~ ~~IIII~~ ~~IIII~~	15
110–164	~~IIII~~ ~~IIII~~	10
165–219	III	3
220–274	II	2
275–330	I	1

The observed frequency is then plotted on the y-axis of the histogram and the classes are plotted on the x-axis (see figure 4.11). The rectangles which result should touch each other, thus reflecting the continuous nature of the observations.

Population pyramids or age-sex pyramids

Population pyramids are a form of histogram used to show the number or, more commonly, the percentage of people in different age groups of a population. They also illustrate the female–male composition of that population. Figure 4.13 is an example of a population pyramid.

Construction of a population pyramid

Although a population pyramid is a form of histogram, a few peculiarities do bear noting. As figure 4.13 illustrates, a population pyramid is drawn on one vertical axis and two horizontal axes. The vertical axis represents age and is usually subdivided into five-year

Figure 4.13: Example of a population pyramid. Australian resident population, 30 June 1993

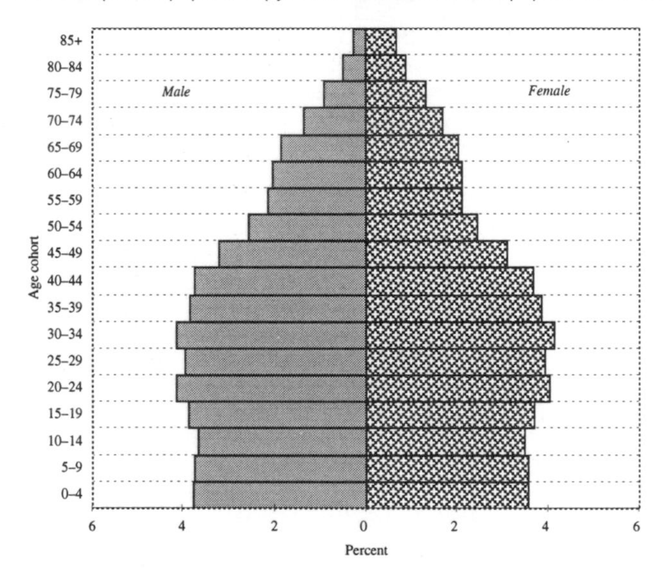

Data Source: ABS (1994a, p.2)

age cohorts (e.g. 0–4, 5–9, 19–14 years). The size of those cohorts may be changed (e.g. to 0–9, 10–19, 0–14, 15–19) depending on the nature of the raw data and the purpose of your pyramid. Remember from the discussion of histograms, however, that if you depict different size cohorts in the same pyramid, the area of each bar must reflect that variability. For example, a 0–14 cohort would be half as wide as a 15–44 cohort on the same graph. On either side of the central vertical axis are the two horizontal axes. That on the left of the pyramid shows the percentage (or number) of males while that on the right shows the figures for females. You will also see from figure 4.13 that the zero point for each of the horizontal axes is in the centre of the graph. As a final note, it may also be helpful for your reader if you include a statement within the graph of the total population depicted.

Circle or pie charts

Pie charts show how a whole is divided up into parts and what share or percentage belongs to each part. Pie charts are a dramatic way of illustrating the relative sizes of portions of some complete entity (Windschuttle & Windschuttle 1988, pp. 272–273). For example, a

Figure 4.14: Example of a pie chart. Total delivered energy, South Australia, 1990–1991 (Total = 189 PJ)

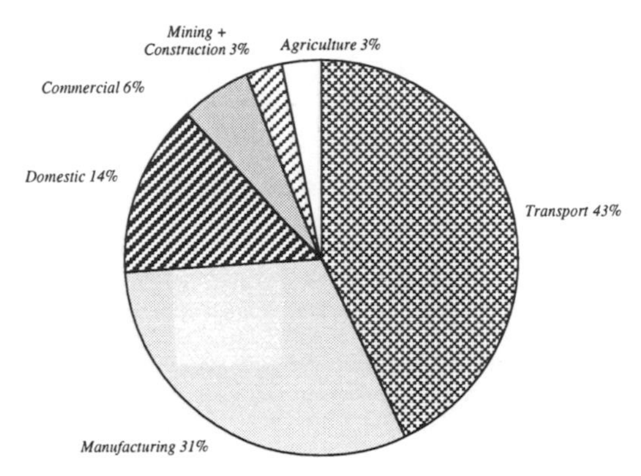

Source: Department of Environment & Land Management (1993, p.207)

pie chart might show how a budget is divided up or who receives what share of some total. See figure 4.14 for an example.

Construction of a pie chart

Constructing a pie chart usually requires a little arithmetic. It is necessary to match the 360 degrees which make up the circumference of a circle with the percentage size of each of the variables to be graphed. Simply, this is achieved by multiplying the percentage size of each variable by 3.6 to find the number of degrees to which it equates. Obviously, if the values have not already been translated into percentages of the whole this will need to be done first.

For example, at the 1981 census there were 1 319 327 people in South Australia and 14 926 786 in the whole of Australia (i.e. the South Australian population was 8.8% of the national total). If a pie diagram of the Australian population by states and territories were to be drawn, the segment representing South Australia would have an angle of:

(1 319 327 / 14 926 786) x 360 = 31.8 degrees.

It is best not to have too many categories (or 'slices') in a pie chart as this creates visual confusion. Five or six segments would seem to be a fair maximum. Generally, no segment should be smaller than six degrees. This may require that some classes be grouped together.

The sectors in a pie chart normally run clockwise, with the largest sector occurring first (ABS 1994g, p. 117). The starting point for the first sector is created by drawing a vertical line from the centre of the circle to the 12 o'clock position on the circumference (Jennings 1990, p. 19).

Pie charts should also advise the reader of the *total value* of categories plotted, as shown, for example, by the statement in figure 4.14 of the total amount of energy delivered. There is little point in letting a reader know percentages without allowing them the opportunity to determine exactly how much that percentage represents in absolute terms.

Logarithmic graphs

Logarithmic graphs are used primarily when the range in data values to be plotted is too great to depict on a graph with arithmetic axes (commonly a scattergram or line graph). Comparative national Gross Domestic Product figures are good examples of such data, for national figures range from millions of dollars to billions and trillions of dollars. Similarly, historical population figures, which might grow from hundreds to thousands to millions, sometimes necessitate the use of a logarithmic graph. Figures 4.15a and 4.16 are examples of logarithmic graphs.

Logarithmic graphs are sometimes also used to compare rates of change within and between data sets. Despite vast differences in numbers, if line *slopes* in a logarithmic graph are the same, then the *rate* of change is similar. This might be useful, for example, if one was illustrating historical rates of population change in a region and trying to argue that despite the fact that the population is now growing at millions of people per year the rate of change has not actually altered since the late 1800s, when the population was growing by thousands each year (see figures 4.15a and 4.15b for example).

Before discussing this form of graph any further, it might be helpful to say a little about logarithms. The logarithm of a number is the power to which 10 must be raised to give that number. For example, the log of 100 is 2 because $10^2 = 100$ (i.e. 10 raised to the power 2 = 100). Thus, the log of 10 is 1; the log of 100 is 2; the log of 1000 is 3 and so on.

Second, simple line graphs and scattergrams, as described earlier, typically use an *arithmetic* scale on their axes (e.g. 1, 2, 3, 4 or 0, 2, 4, 6...) where a constant numerical difference is shown by an equal interval on the graph axes. In contrast, semi-logarithmic and logarithmic graphs use a *logarithmic* scale where the numerical value of each key interval on the graph increases *exponentially* (e.g. 10, 100, 1000, 10 000)

Figure 4.15: Comparison of data displayed on (a) semi-log and (b) arithmetic graph paper

(a) South Australia's resident population, 1884–1991

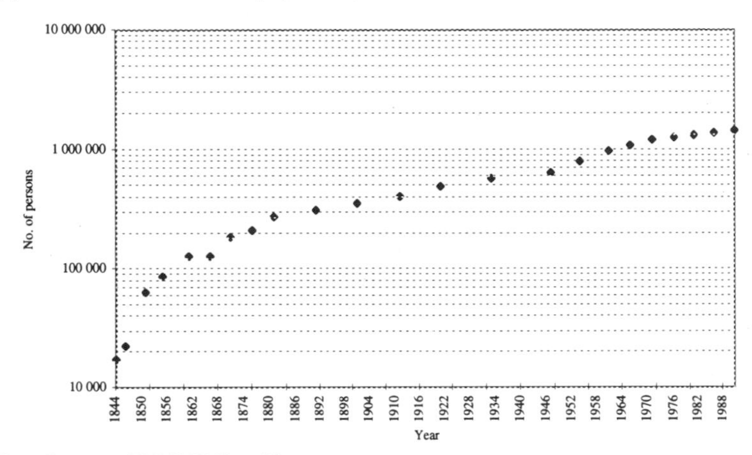

Data Source: ABS (1994f, p.49)

(b) South Australia's resident population, 1884–1991

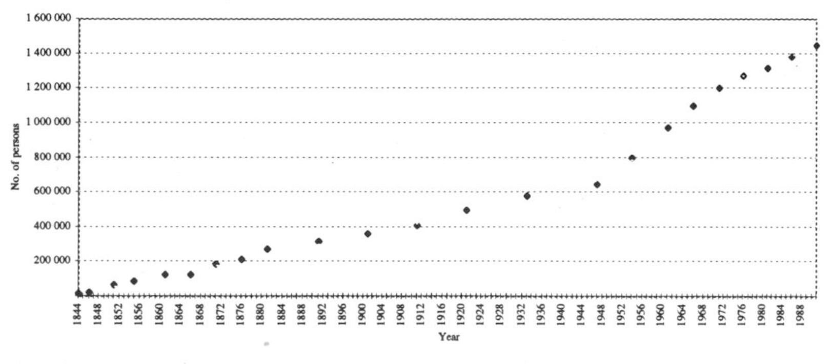

Data Source: ABS (1994f, p.49)

and the lines in each *cycle* (each cycle is an exponent of ten) of the graph become progressively closer together (see figures 4.15 and 4.16 for examples). Figure 4.15a is an example of a *semi-logarithmic* graph. It has a logarithmic y-axis and an arithmetic x-axis. (Histograms and column graphs can also be drawn using semi-logarithmic paper if the variable to be depicted through the y-axis has a particularly large range.) Figure 4.16 is a *log-log* graph—both axes are logarithmic.

Figure 4.16: Example of a graph with two logarithmic axes (a 'log-log graph'). Relationship between gross domestic product and passenger cars in use, selected countries, 1992

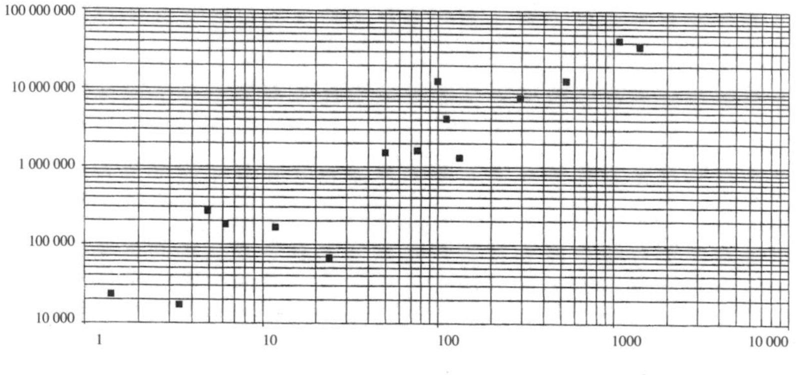

Data Source: World Almanac 1995

Zero is never used on a logarithmic scale because the logarithm of zero is not defined. Scales on log paper start with ... 0.001, 0.01, 0.1, 1, 10, 100, 1000, 10 000, 100 000 ... (or any other exponent of ten). If you look at figure 4.16 as an example, you will see that the x-axis commences at 1, whereas the y-axis commences at 10 000. The determination about the figure with which to start the axis is made on the basis of the smallest data point. For example, if the smallest figure for plotting was 35 000, you would start the axis at 10 000, not 1000 or 100 000.

Construction of a logarithmic graph

Consider the maximum and minimum values of the data sets to be plotted. In looking, for example, at the population of the state of South Australia over the period 1844–1991, we see that the population grew from 17 366 people to 1 446 229 people. To accommodate this range of data the graph we will need to have three logarithmic cycles, with the first commencing at 10 000, the second at 100 000 and the third at 1 000 000. Because the years are plotted arithmetically, the graph will be drawn in semi-log format. Simply plot data points at their appropriate (x,y) coordinates. Add an appropriate title and indication of data source. The same procedures apply for log-log graphs except, of course, that it is necessary to consider the number of cycles for both axes, not just one.

Tables

Tables present related facts or observations in an orderly, unified manner. They are used most commonly for summarising results. Tables can be effective for organising and communicating large amounts of information, especially numerical data, although the mistake of trying to communicate too much information at once is commonly made.

The main reasons for using tables are to facilitate comparisons, reveal relationships, and save space. Tables should be constructed so that they are comprehensible without having to refer to additional explanation in the text (Campbell, Ballou & Slade 1986, p. 152). The data in the table may be referred to and discussed but it should not be repeated extensively in the text. Table 4.4 is an example of a correctly set out table.

Table 4.4: Example of a table. Major groundwater resources of Australian states/territories, 1987

		Ground water resource (gigalitres)				
State/ Territory	*Area of aquifers (km²)*	*Fresh*	*Marginal*	*Brackish*	*Saline*	*Total*
New South Wales	595 900	881	564	431	304	2180
Victoria[1]	103 700	469	294	691	30	862
Queensland	1 174 800	1760	683	255	144	2840
South Australia	486 100	102	647	375	86	1210
Western Australia	2 622 000	578[2]	1240	652	261	2740
Tasmania	7240	47	69	8	–	124
Northern Territory	236 700	994	3380	43	10	4420
Australia	**5 226 440**	**4831**	**6877**	**1833**	**835**	**14 376**

Notes:

1 In case you had not noticed, I inserted a note identifier with the entry for Victoria simply to illustrate how a table footnote might look. See the discussion below to find out about the purpose of such notes.

2 Look, I did it again—this time beside the freshwater data entry for Western Australia.

Source: Australian Bureau of Statistics (1994a, p. 18).

Elements of a table

Aside from the information being conveyed, the main elements of a table are:

i *Table Number*—each table should have a unique number (e.g. table 1) allowing it to be easily identified in textual discussion.

ii *Table Title*—the title, which is placed one line above the table itself, should be brief and allow any reader to fully comprehend the information presented without reference to other text. The title should answer 'what', 'where', and 'when' questions.

iii *Column Headings*—headings are necessary to explain the meaning of data appearing in the columns. It is a good idea to specify the units of measurement (for example $, mm, litres) within the column headings (only a small amount of space is available for headings so they must be concise, however not so concise that they become ambiguous). Below the column headings a dividing line is placed to separate them from the data. The bottom of the table should be marked with a single horizontal line.

iv *Table Footnotes*—footnotes appear below the table to provide supplementary information to the reader, such as restrictions that apply to some of the reported data. Table footnotes may also be used for explanation of any unusual abbreviations or symbols.

v *Table Source*—an indication of the source from which the data was derived or the place from which the table was reproduced should be provided. An accepted form of referencing must be used. See the chapter on referencing for more information.

Maps

> 'A map is often the heart, or better, the brain, of a scientific paper' (Paraphrase of Morgan, in Day 1989, p. 56).

'A map is a graphic device used to show where something is' (Moorhouse 1974, p. 86). Maps use labels, symbols, patterns, and colours to convey messages about spatial and other relationships. Maps are marvellous for exploring research questions and for pointing out relationships which might otherwise be difficult to see. Indeed, maps can be so intellectually provocative that they raise more research questions than they resolve!

The following sections of this chapter comprise brief discussions on the character and preparation of three common types of statistical map (i.e. ones which show spatial variation in amounts and quantities). These are dot maps, choropleth maps and isoline maps. The specific type of map you might choose to use will depend on the nature of your data and the purpose you have for your map.

Dot maps

The dot map is a common way of showing both the spatial distribution and the quantity of a variable. Figure 4.17 is an example of a

Figure 4.17: Example of a dot map. Distribution of Aboriginal population, Adelaide, 1981

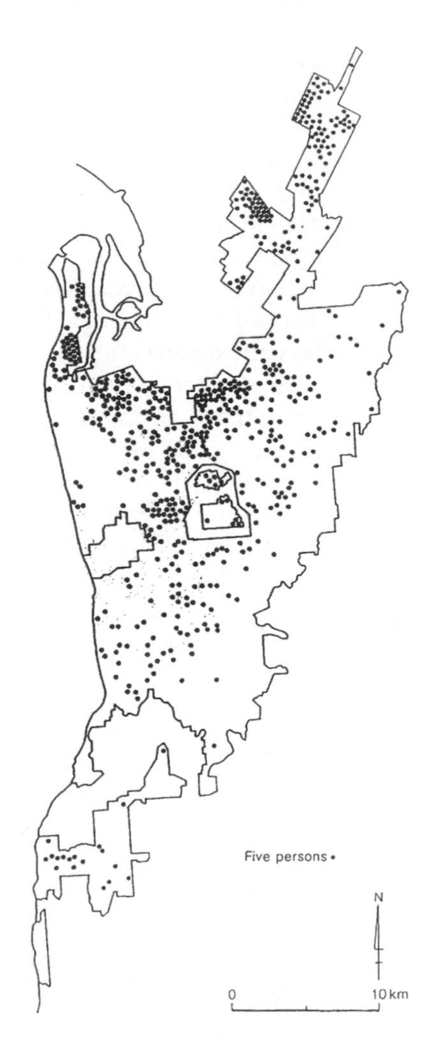

dot map. Simply, a dot representing every occurrence of a given characteristic is placed on the map (Toyne & Newby 1971, p. 91). Dots are useful for showing the distribution of discontinuous or discrete data sets (e.g. population, stock numbers). Each dot may represent some multiple of the individual units being depicted. For example, a single dot might represent 1000 people or 300 sheep.

Construction of a dot map

i Decide on the number of units each dot will represent. The scale of dots needs to be chosen carefully if it is to be effective. If the value of each dot is too large, sparsely populated districts may not be represented at all. If the value is too small, dots will join in densely populated districts (Garnier 1966, p. 16) and you are likely to go crazy drawing the map. The size of the actual dot is also important. If too large, it will make the map look messy and crude. If too small, it will fail to depict spatial variation. Try plotting the greatest and lowest densities to be shown on the map (see the next step for some help on that) using several different dot values to get some impression of the value that will be most effective.

ii Dot maps are generally divided into areas such as statistical divisions or administrative units, so for each division, calculate the number of dots to be shown and pencil in those numbers on your map.

iii Draw the appropriate number of dots, ensuring that they are spaced evenly (but not in lines). However, if variations are known to occur within the area, attempt to locate the dots in ways which best represent the spatial variations you know about. For instance, in drawing a dot map of the population distribution of Australia or New Zealand it would be inappropriate to distribute dots evenly across the entire country. Instead, dots would be placed in focal areas near major population centres.

iv Complete the map by adding a title, legend, scale and source of data.

Proportional circle maps

Proportional circle maps are a variation on the dot map. Instead of using dots of identical size to represent numbers of some phenomenon, the size of each circle in a proportional circle map is, as the name suggests, directly related to the frequency or magnitude of the phenomenon represented. Figure 4.18 is an example of a proportional circle map.

Construction of proportional circles

Circles may illustrate quantities on maps by using a scale which is related to either:

i the *diameter* of a circle, or

ii the *area* of a circle.

Figure 4.18: Example of a proportional circle map. Distribution of Vietnamese-born persons, Adelaide Statistical Division, 1986

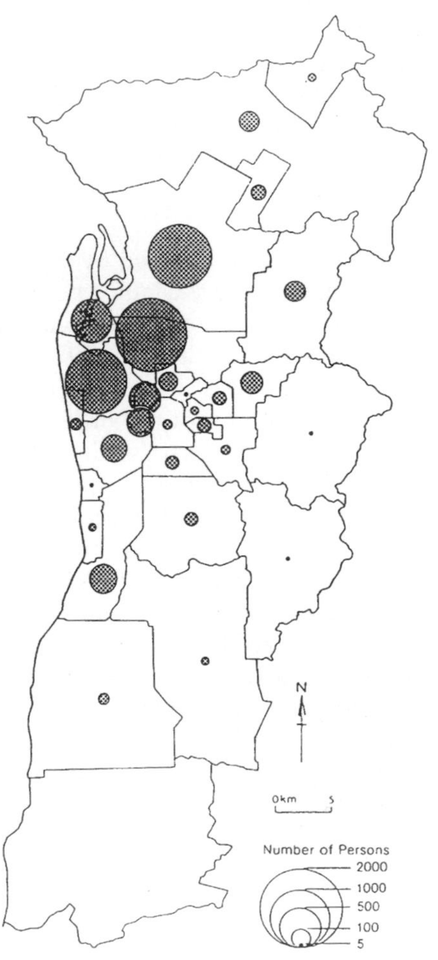

Both methods are discussed below. The area-based representation of quantities is to be preferred as it provides a more accurate visual portrayal of quantity than does the diameter-based approach.

Proportional circles—diameter-based

If the populations of two small towns (Port Gerard with 400 people and Susanville with 1600 people) are plotted using this form of proportional circle, the *diameter* of the circle representing the second town should be four times as large as that of the circle representing the first.

The diameter of the circle is directly proportional to the given value. Thus, if you have decided that each millimetre of circle radius will represent 100 people, the circle diameter for Port Gerard will be 4 mm whilst that of Susanville will be 16 mm (Garnier 1966, p. 18). This is a simpler method of drawing proportional circles than the area-based technique, but it overemphasises the visual impact of large values because circle areas grow exponentially with increases in radius.

Proportional circles—area-based

In this method, statistics are portrayed by circles whose *areas* are proportional to the size of the variables being depicted. This is achieved by making the diameter of the circle proportional to the square root of the number being illustrated. Taking the example discussed above, the area of the larger circle should be four times that of the smaller one. The square root of Port Gerard's population of 400 is 20 and the diameter of that circle will be proportional to that number. The square root of Susanville's population of 1600 is 40 and, following either the graphical or mathematical method of calculating circle size outlined below, an appropriately sized circle will be drawn. Even though the square root of Susanville's population is only two times as large as that of Port Gerard, the circle area will be four times greater thanks to the magic of mathematics. The end result will be two circles whose areas give an accurate visual representation of the fact that one town has four times as many people as the other.

To create area-based proportional circles using a graphic method, follow the steps listed below.

i Decide on the radius of the biggest circle that can be used on your map or figure. This will depend on the scale of the map and the number of proportional circles to be shown. Aim for a result which is a happy medium between a small number of insignificant circles and something that looks like a bubble-bath.

ii Calculate the square root of each of the quantities to be illustrated.

iii Construct a continuous scale of circle size from which you can read off the radius of any of the circles to be drawn. (It may be useful to consult figure 4.19 as you read this section and the next.) To make this scale, find yourself a spare sheet of paper and create a horizontal arithmetic scale which:

- is divided into equal units;
- begins at zero; and
- embraces the whole range of the square roots you calculated in the first step.

Figure 4.19: Example of a graphic scale for creating proportional circles

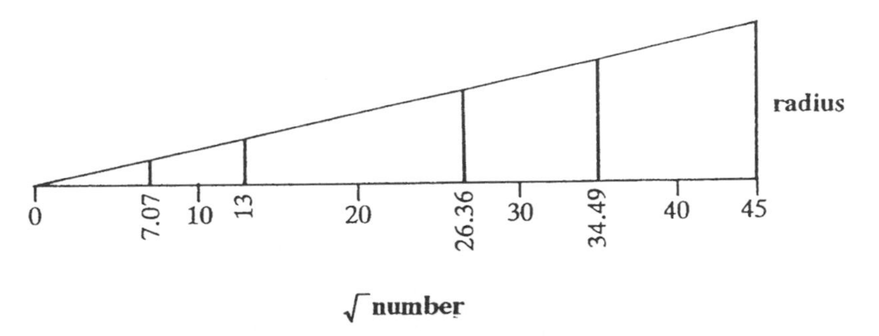

iv In step i you decided what the maximum circle radius should be. Now draw a vertical line of that length upwards from your horizontal arithmetic scale at the point corresponding with the largest square root value you calculated in step ii. Join the top of that vertical line to the zero point on the horizontal axis. Believe it or not, the result should be a triangle which will allow you to read off required radii from the horizontal axis without any further calculations. See figure 4.19 for an example.

v To use the scale you have created, place the point of your compass on that part of the horizontal axis which corresponds to the square root of the value to be represented. Open the other leg of the compass to reach that point of the diagonal line directly above the square root value to be plotted. With the compass set at this radius, draw a circle on the map (or illustration) centred on the location of the area being plotted.

vi A circular scale must be shown on the finished map. A neat and simple way of doing this is to draw circles corresponding to rounded representative figures in the data (not the square root values). See figure 4.20 for an example.

Figure 4.20: Example of a completed proportional circle scale

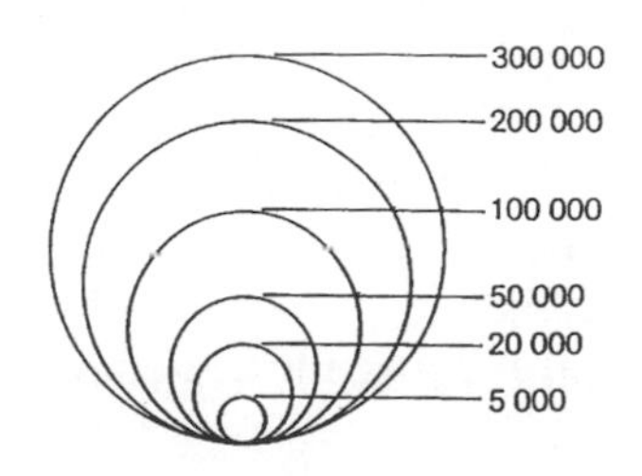

Do *not* use the triangular scale you created to construct the proportional circles as the scale in your illustration.

Following International Cartographic Association (1984, p. 106) advice, proportional circle areas can also be calculated mathematically. This is a relatively straightforward procedure.

Say,

N = maximum value to be represented
n = one of the other values in the data series
S = the area of the circle representing N
s = the area of the circle representing n
R = the radius of the circle representing N
r = the radius of the circle representing n

then $\frac{s}{S} = \frac{n}{N}$ or $\frac{\pi r^2}{\pi R^2} = \frac{n}{N}$ thus $\mathbf{r} = \frac{\mathbf{R}\sqrt{\mathbf{n}}}{\sqrt{\mathbf{N}}}$

You will see that this procedure still requires you to take steps i and ii outlined in the graphic method of calculation. Of course, you will also need to prepare a scale to put on the completed map (step vi above).

The technique of using proportional circles may be usefully extended to represent two quantities (e.g. freight tonnages in and out of a region) in the form of a *split proportional circle*. To achieve this, follow the instructions above, but draw two semi-circles at each point. Figure 4.21 provides an example.

Figure 4.21: Example of a split proportional circle

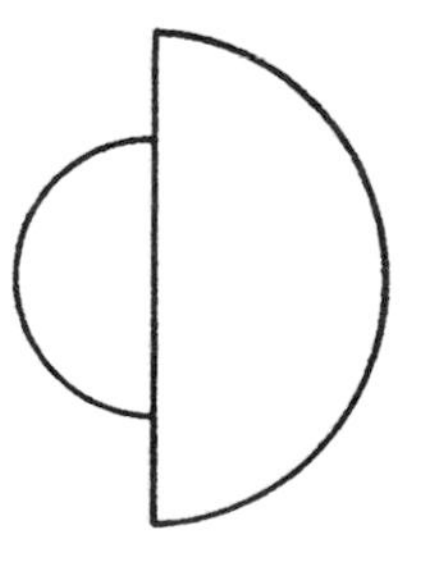

The semi-circle on one side might, for example, represent freight into a region or births in a place, whereas the semi-circle on the other side would represent freight exports or deaths. The circles are made proportional to the figures being illustrated in exactly the same way as discussed above, but only half of the circle is actually drawn. The same scale must be used for the quantities in each half or comparison would be impossible.

Another way of displaying information on maps is as a *proportional pie diagram*, where the size of the circle will show the magnitude of a given value. However the circle is then subdivided into different size segments, each of which represents some percentage of the total value (see the discussion on pie charts for more information). For example, a map of housing availability in Darwin might show proportional circles indicating the number of homes in each suburb, and these in turn might be internally subdivided to show the proportions of government housing, owner-occupied housing, rental dwellings and so on in each of the mapped suburbs.

Choropleth maps

A choropleth map displays spatial distributions by means of cross-hatching or shading. Figure 4.22 is an example. Choropleth maps are commonly used to display rates, frequencies and ratios such as marriage and divorce rates, population densities, birth and death rates, percentage of total population classified according to sex, age, ethnicity and per capita income. Choropleth maps may also be used to show non-statistical data such as soil types, vegetation types (Moorhouse 1974, p. 92).

Construction of a choropleth map

The description that follows outlines the preparation of a choropleth map depicting statistical information. Non-statistical maps require completion of only the last three steps.

Preparing a choropleth map involves a number of basic steps (Sullivan 1993, pp. 69–71).

i Calculate the range of the data to be mapped. Remember, the range is the difference between the highest and lowest value. For example, if the highest value is 250 mm and the lowest is 72 mm then the range is 178 mm (i.e. 250 - 72 = 178).

ii Decide on the number of classes into which the data will be grouped. This will depend on the purpose of the map and the nature of the data. If there are too many classes the values for specific areas may be difficult to identify. Four to six classes is generally sufficient to identify spatial patterns without making the map too detailed.

Figure 4.22: Example of a choropleth map. Women in paid employment, Adelaide Statistical Division, 1991

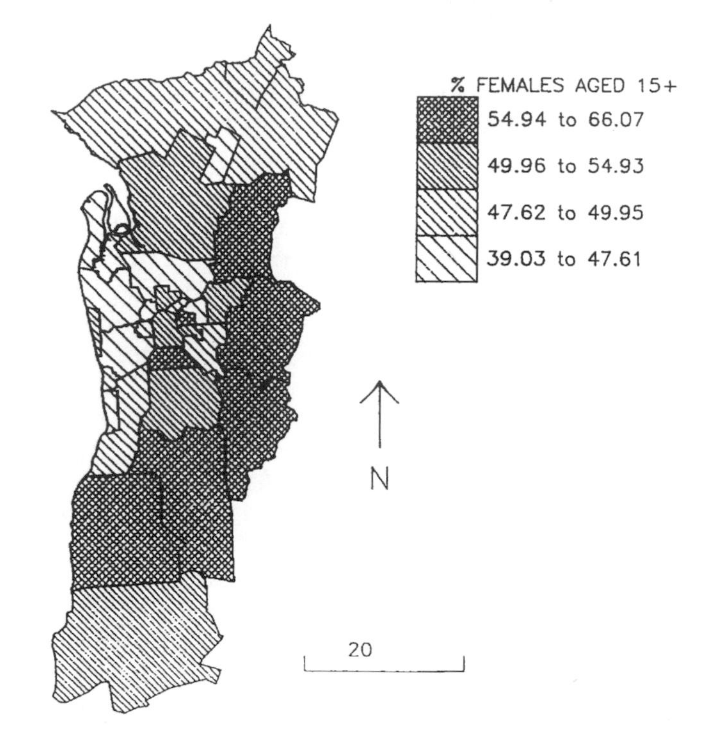

Source: ABS Census 1991

iii Determine the interval or range of values within a single class. To do this, follow the directions outlined in the earlier discussions on class intervals and frequency distributions under 'Construction of a histogram'. In general, the outcome should be one in which there are approximately equal size classes (in terms of class interval or number of observations in each class) and there should be no vacant classes (Toyne & Newby 1971, p. 86).

iv Create a cartographic pattern (grey tones or colour) to represent each of the class intervals you have selected. This will be your *legend*. Low values are typically represented by light colours or shading and high values by darker colours or shading. An easy method is to use a single colour changing progressively from light to dark. Use colours that match the phenomenon being mapped.

For example, greens for vegetation, browns for soils. See a good atlas for examples.

v Transfer the cartographic patterns of the legend to the map. For every geographic unit such as a nation, state, region or shire, shade or colour that unit according to the pattern in the legend. Avoid having blank areas, which do not provide any information.

vi Complete the map by adding a title, scale, northpoint etc.

Isoline maps

Most of us see one form of isoline map every day when we look at newspaper and television weather reports. In common with other forms of isoline map, weather maps show sets of lines (isolines) con-

Figure 4.23: Example of an isoline map. Average annual rainfall, South Australia (isohyets in millimetres)

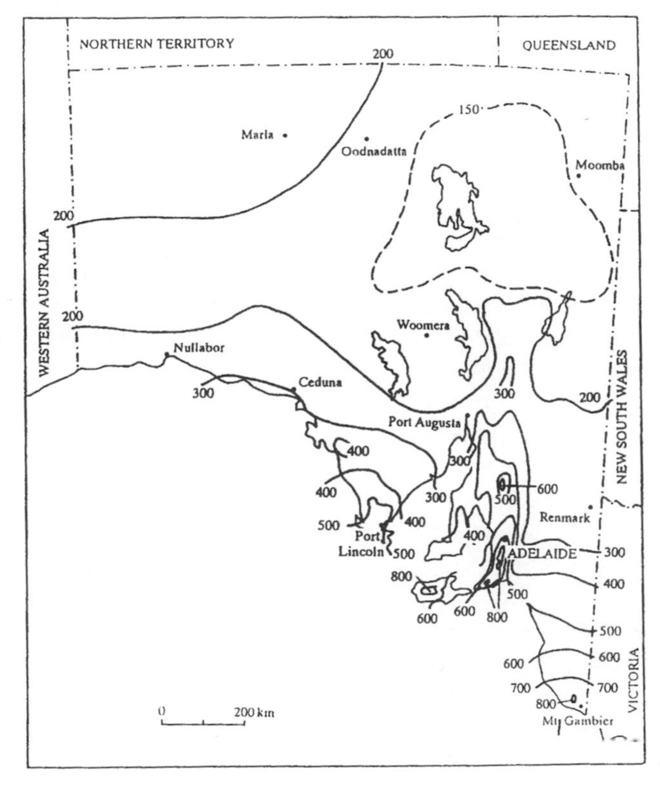

Source: ABS (1994f, p. 5) Copyright in ABS data resides with the Commonwealth of Australia. Used with permission.

necting points known, or estimated to have, equal value (atmospheric pressure is usually portrayed in weather maps). See figure 4.23 for an example. The topographic map, depicting contours of equal elevation, is another frequently encountered form of isoline map.

Isoline maps always use data with continuous distributions (e.g. rainfall, temperature). Table 4.5 lists common types of isoline and the variables they depict.

Table 4.5: Some common isolines and the variables they depict

Isoline	*Connecting Places of Equal*
Isobar	Atmospheric pressure
Isotherm	Temperature
Isohyet	Rainfall
Contour line	Elevation
Isobath	Water depth

Construction of an isoline map

The following steps are required to prepare an isoline map (based on Garnier 1966).

i On a base map of the area being represented, locate all points for which precise figures for the phenomenon being mapped are available. See figure 4.24(a).

ii Calculate the range of the values being mapped and, taking that figure into consideration, decide on a suitable value for the interval between each isoline (i.e. for a contour map, should it be 5 metres, 10 metres, or 100 metres ...?). In making this decision, it is helpful to consider the number of known observations upon which you are basing the map. Toyne and Newby (1971, p. 99) suggest that the number of isolines (classes) be no more than five times the logarithm of the number of observations. Thus, for example, if your map has 100 observed points on it, one might expect a map with ten isoline classes. A map with 23 known points would have about seven isoline classes. So, if you are drawing a topographic map on the basis of 100 observed points with a data range of 900 metres (say from 800 m to 1700 m), it would seem appropriate to aim for an interval between each isoline of about 100 metres (e.g. 800 m, 900 m, 1000 m ...). Remember, however, that the interval you choose should be set at a value that will show the detail of the distribution without overcrowding the map. It may be helpful to do several trial plots of the data to determine the best interval to use.

Figure 4.24: Three steps to creating an isoline map

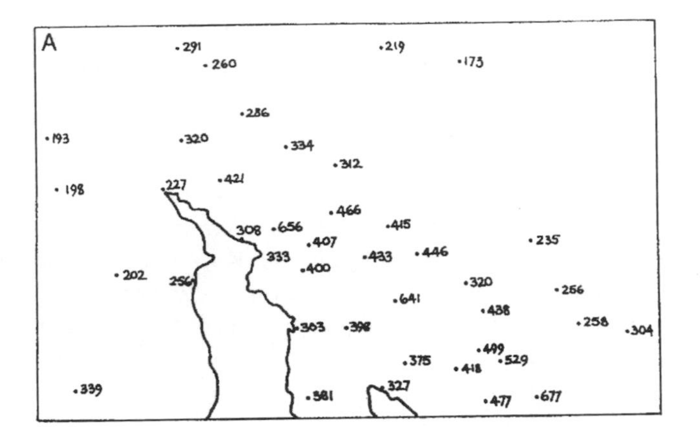

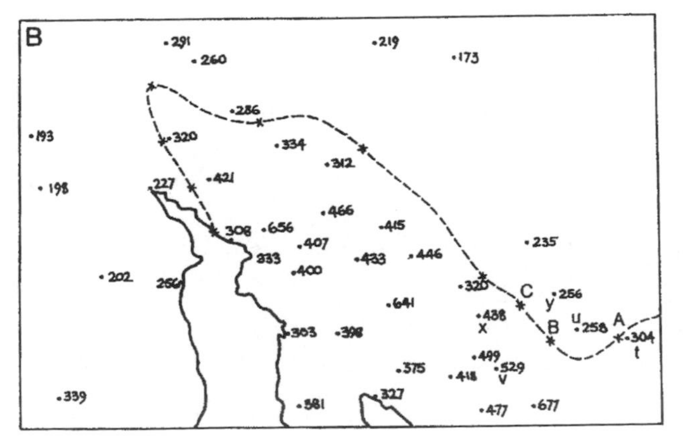

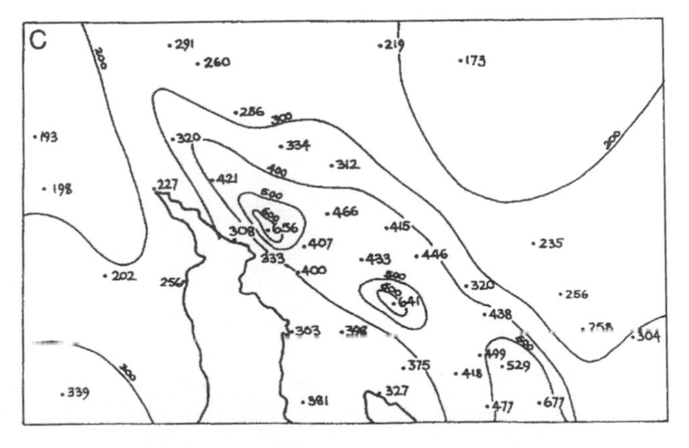

iii Draw the isolines. This is the most difficult part. Disappointing as it may be, the process is not a case of just joining dots. Most point data will not correspond exactly to the contour values decided in the preceding step. The position of isolines *between* point data of different values is worked out by the technique of interpolation. Figure 4.24(b) shows how this is done for a map of precipitation in South Australia. Assume that the rate of change between one known point and another is constant, unless information you have suggests otherwise. Point *A*, which depicts isohyet value 300, is between point *u* (258 mm) and point *t* (304 mm); point *B* lies about one-quarter of the way between points *u* (258 mm) and *v* (529 mm); and point *C* is somewhat closer to point *y* (256) than it is to point *x* (438 mm).

iv The appropriate points are then joined by smooth lines as shown in figure 4.24(c). Make sure that your map has enough isolines to show the data pattern accurately without creating visual confusion. If you have too many isolines, or if the data pattern is not depicted as well as you would like it to be, rethink your selection of isoline interval and redraft the map.

v Label enough of the isolines with their values to enable readers to understand the map quickly. Place the isoline values so they can be read without turning the map.

vi Complete the map by adding a title, scale, northpoint etc.

5

Preparing posters

The results of research and other scholarly enterprise are not conveyed only in the form of talks, essays, or research reports. Posters are an alternative and effective means of presenting an idea or set of ideas. In part recognition of this, posters are becoming an increasingly important medium of scientific and professional communication. Posters can present a real challenge; they require careful combination of effective graphic and written communication. This chapter outlines some of the keys to poster production. Discussion covers layout, visibility, and the use of colour and type style. While there is also a brief reference to the use of figures, an extended discussion on the production of graphic devices can be found in the previous chapter. The chapter concludes with a poster assessment schedule.

Why Prepare a Poster?

Posters are especially good for promoting informal discussion, for summarising a project, and showing results that require more time for interpretation than is possible in, say, an oral presentation. However, they are not useful for reviewing past research or presenting the results of textual research (Lethbridge 1991, p. 14).

A poster presents an argument or explanation or outlines the results of some piece of research. Physically, a poster is a piece of stiff card measuring about 90 x 60 cm (some may be much larger) to which graphic materials such as maps, graphs, and photos are affixed and linked together by a small amount of text. Figures 5.1 and 5.2 show examples of posters completed by two first year university students studying geography in the USA.

Posters are assigned for a variety of reasons. The first posters I ever asked students to complete were part of a large-scale teaching strategy I shared with some colleagues and which we believed would add

Figure 5.1: Example of a poster completed by a first year university student

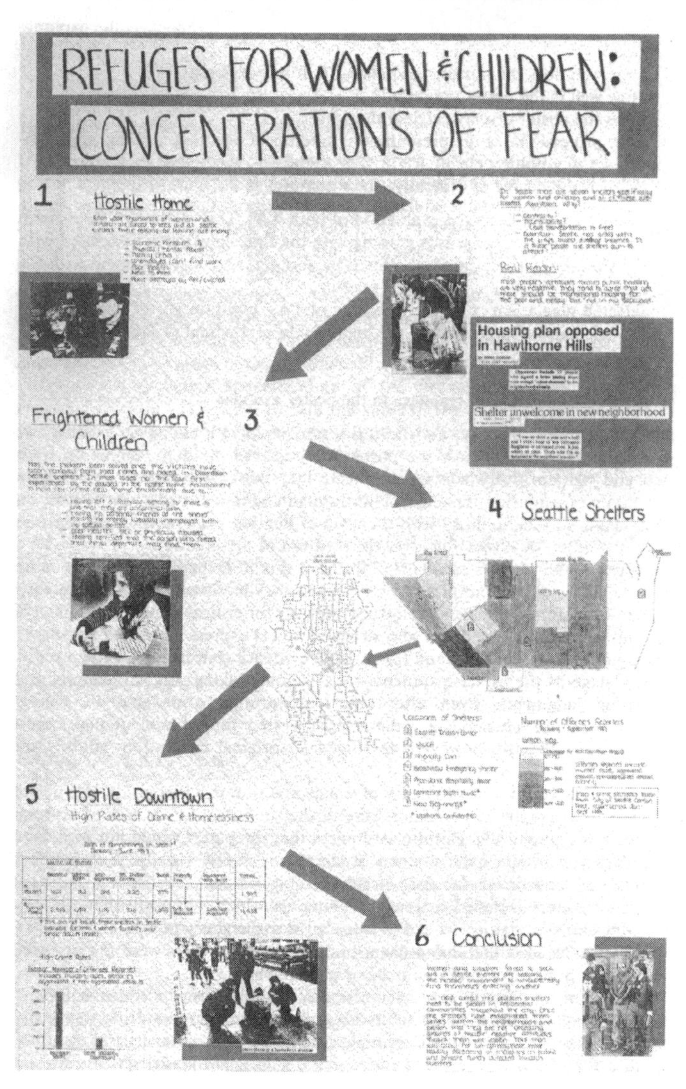

variety and new challenges to the course we taught; stimulate critical thought; and encourage student-staff interaction (see Howenstine et al. 1988 for a discussion). Your lecturer may have given more thought to the decision and might be seeking to help develop your skills in graphic communication. S/he may have recognised that in an age of symbols and sight, posters are a particularly useful form of commun-

Figure 5.2: Example of a poster completed by a first year university student

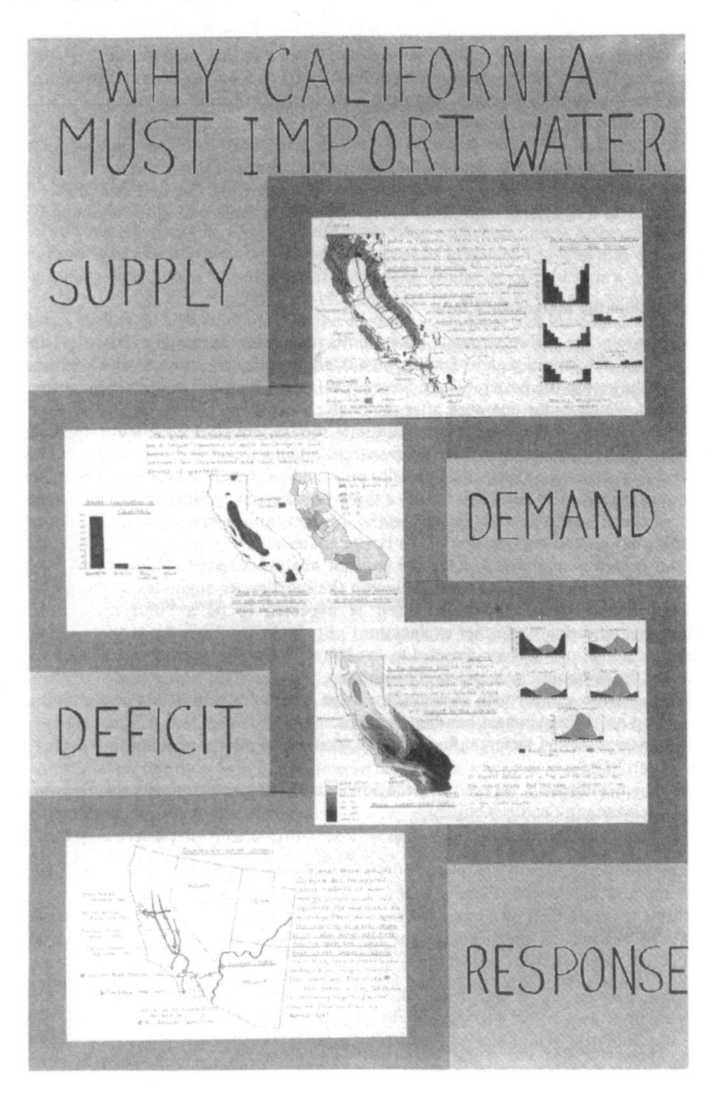

ication. S/he might also be acknowledging an understanding that increasing numbers of academic and professional gatherings now solicit submissions in poster form (Hay & Miller 1992, p. 202; Committee on Scientific Writing, RMIT 1993, p. 58; Day 1989, p. 143).

If you are assigned a poster exercise, do not make the mistake of thinking that it involves little more than crayons, colouring-in and collage. Most students find posters a particularly challenging means of communicating information. Posters require you to express complex ideas with brevity and grace and to balance text with high quality graphics.

What are Your Poster Markers Looking For?

A few central principles are critical to the creation of a successful poster. It is performance with respect to these principles that your markers are looking for. The five principles might be remembered as the A, B and C of poster production:

- **Attention-getting:** does the poster make a good first impression? Does it grab the attention of the viewer? Achieve this through layout, colour, title, and other devices.
- **Brevity:** the poster should make its point(s) quickly.
- **Coherence:** an effective poster makes a logical, unified statement requiring no further explanation. It should be intellectually accessible to the intended audience and must be capable of 'standing alone.'
- **Direction:** simplicity would have been a better word, but it did not fit in with the A, B, C mnemonic! Over-complicated posters discourage and confuse readers. Keep the poster simple. Keep it focused.
- **Evidence:** your argument must be supported by accurate, referenced evidence.

Not surprisingly, these principles are very similar to the keys that unlock successful written and oral communication. The principles outlined above are expressed in rather more detail in the poster assessment schedule at the end of this chapter.

The A, B, Cs of poster production underpin virtually all academic posters. To these general principles some others, which reflect specific requirements of particular types of poster assignment, might be added. For an academic poster you will probably be required to:

- express a problem statement and resolve the problem;
- *argue* or *explain* an issue; and/or
- *evaluate* evidence concerning a chosen topic (Howenstine et al. 1988, p. 144).

These require you to engage in critical thinking rather than simply to describe some phenomenon. You will notice that these matters are covered in the poster assessment schedule. If you are given free choice of poster topic by your lecturer, you would be wise to check that your approach to the selected topic is appropriate before you proceed too far. Make sure that you are satisfying the intellectual demands of the exercise as well as the graphic requirements.

Text

Number of words

Unlike posters advertising airlines and alcohol, academic posters almost always contain some amount of text. Together with the graphics, text contributes to the introduction, explanation and discussion of your work. It is important however to keep text to a minimum. Lethbridge (1991, p. 18) suggests a maximum word count of 500, but makes the point that a poster will be much more effective if it uses only 250–400 words. Confine the text you do use to short sections that complement the poster's graphic components. Audiences do not enjoy reading long texts. Bad posters are usually bad because they contain too much text or because all the text is presented in one large chunk.

Level of detail

Although a poster should always be self-contained, needing no further explanation, it need not contain all the details of your project. In much the same way as a talk highlights points, which can then be taken up by interested members of the audience, so too a poster can be used to present only the most important elements of your work. Particularly interested onlookers might speak to you about details (Day 1989, p. 145). Having said this, the initial point bears repetition: the poster must still be comprehensible without further elaboration.

Text and reading patterns

Be aware that after reading the title and scanning the poster many people move straight to the conclusions (Simmonds & Reynolds 1989, p. 95). If the conclusions seem important, the viewer may then decide to read the rest of the poster. In light of this, give some hard thought to your title and to the way you present your conclusion(s).

Layout

Layout is an important part of graphic communication. Although a poorly prepared poster is unlikely to send people to sleep, it is also unlikely to attract and hold an audience in the way that your work may deserve. Think carefully about the way you set out your poster.

Layout and research

Because thoughtful composition is such an important part of effective communication (Vujakovic 1995, p. 254), it is worth setting aside a big chunk of time to laying out your poster. It is also a good idea to design your poster at the same time as you are conducting the research associated with the exercise. Produce sketches of poster layout prior to final set-up. You will probably find that this will save you time and that it will help you make sense of the issue being discussed in the poster. For example, in laying out the poster you may find weaknesses and gaps in any argument you are developing or relationships you are exploring. The detection of such gaps, which is often simpler with a poster than with written work, may stimulate additional inquiry, which will contribute to the production of a better research project. The connection between project development and poster layout, plus the difficulty of disguising errors of logic in a completed poster, also make the all-too-common efforts to try to piece something together the night before a poster assignment is due quite inappropriate, no matter how straightforward the project may initially appear.

Elements of a poster

Besides graphics, posters usually comprise five important components:

i title and subheadings which should be visible, short, and memorable;
ii an introduction, in which a short statement of the problem investigated and the approach used are provided;
iii the body of the presentation, containing pertinent information such as a short description of research methods and data used;
iv a statement of conclusions and/or directions for future research;
v references, placed on the front of the poster or, in some cases, attached to the reverse side.

It is clear that in their structure, posters are little different from essays and other forms of expression.

Guiding the reader

Unlike written forms of communication, posters do not have to be set up in a linear form, with readers moving from top left to bottom right. 'Spider' diagrams, perhaps showing factors contributing to urban consolidation, and cyclical diagrams, showing the hydrological cycle, are examples of alternatives to linear structure. While posters do offer this flexibility of presentation, it is important that you provide your readers with a clear sense of direction. To this end, readers must usually be guided through the poster with numbers or arrows. They do not know their way around your poster as you do. Lead them along. Some, but not all, topics may lend themselves to novel or eye-catching components in this respect. For example, one student evaluating the implications of ambulance locations for heart attack survival directed readers through the poster with sketches of arteries instead of the more common arrows.

Informality of layout

In creating balance in a poster, informality should take precedence over formality. For example, instead of setting two equal size sheets of paper on each side of the display board, it is more effective to 'play colour and texture, words and form, against each other to get the desired balance' (Larsgaard 1978, p. 193). If shapes are to be used in the poster presentation, similar but not identical shapes should be used throughout.

Visibility and Colour

All materials on the poster should be clearly legible from a distance of about 1.5 metres. Titles and headings should be visible to viewers several metres away, and to further promote ease of reading, text should employ upper and lower case letters, *not* all capitals WHICH ARE A LOT MORE DIFFICULT TO READ QUICKLY. Confine textual materials to brief statements. Do not write an essay and paste it to the board. No-one will read it.

Type

For posters, suggested type sizes are:

- main headings 96–180 point (27–48.5 mm)
- secondary headings 48–84 point (12.9–25.4 mm)
- sectional headings 24–36 point (5.9–8.7 mm)
- text and captions 14–18 point (3.2–4.6 mm)

Use larger size fonts and/or display faces with short headings and smaller and/or less decorative type for longer titles. Twelve point, a common size for typed essays and reports, is much too small a point size for the body of text. If you find you need to use such a point size to fit all of your text on the poster, the poster probably contains too much text. The solution to this problem is to abbreviate and rewrite the text.

Type is one of the most important aspects of visual design, especially for headings, and you should use a typeface that relates to the subject material. While the body of the text needs to be clear and simple, you may be able to add extra graphic character to the poster through your choice of a suitable type. Typefaces and type styles play important roles in providing interest and contributing to the poster's theme and to clarity. When choosing the type to be used on a poster, consider the following elements:

- **typeface:** the particular character of the letter forms, from which there are thousands to choose (e.g. Helvetica, Times)
- **type weight:** the thickness of the letter stroke (e.g. light, regular, medium, bold, extra bold)
- **type style:** which may be roman (upright), italic (slanted), condensed or extended
- **type size**
- **type colour**

For example, a computer space-age typeface might be appropriate for a poster investigating some of the effects of technology, or a playbill typeface first used in the Victorian era on posters advertising stage-productions could be appropriate for a poster considering nineteenth century Australian health care conditions.

Colour

Often one of the most striking and emotive elements of a completed poster is colour, a component which adds to or detracts from the overall impact of the project. Colour can command attention, bring pleasure, and clarify a point (Larsgaard 1978, p. 193). It can highlight important dimensions of a poster or suppress less important facets.

Because colour is important, it should be used judiciously. To avoid confusion and chaos in the poster use as few colours as possible. Take care in your choice of combinations of text and background colours. The text should always be contrasted sufficiently with the background to allow easy reading from a distance. For example, orange text on a yellow background can be difficult to read.

Colour can add symbolic connotations and feelings to the message of a poster. You might find the following list useful. It summarises some colour-connotation connections from an Anglo-American perspective:

- **black:** clearcut and crisp, death, dignity, doom and gloom, financial credit, formal things
- **blue:** calm, climatological, coastal, coolness, rivers, peace, sadness
- **brown:** dismal, dreary, earth, pollution, soils
- **green:** agriculture, conservation, coolness, envy, freshness, growth, nature, rural, safety, spring, vegetation, wealth
- **orange:** autumn, flames healthiness, sunshine, warmth.
- **red:** action, blood, danger, financial deficit, fire, hazards, health, heat, Marxism, noise, passion
- **white:** cleanliness, glory, iciness, purity, snow
- **yellow:** beaches, happiness, light-heartedness, sand, sunshine, weakness

Adapted from Sim (1981)

Posters on cardiac care and on battered women have effectively employed the colour red, but one poster on the subject of women and AIDS is particularly memorable. Black and white were deliberately and dramatically employed as the only colours in the poster. Simply through the use of colour, the stark final product immediately conjured up physical images of life and death while simultaneously raising the spectre of the moral issues of right and wrong which continue to surround the disease.

Particular colours are often associated with specific cultures and nations. For example, red, black and white are often associated with the Maori population of New Zealand and the Native Americans of the US Pacific North-west; red, yellow and black with the Aboriginal population of Australia; and red, white and blue with France, the United Kingdom, and the USA.

Colour may also be used to add information to particular graphics. For example, the appropriate use of red and black conveys the message that financial results portrayed graphically represent profit and loss statistics, not simply dollar figures. In sum then, colour symbolism and association can significantly affect the impact and effectiveness of a poster.

Figures and Photos

Images are a particularly important part of a poster. For example, a photograph can be used to describe what a braided river in New Zealand's Canterbury Plains looks like, while an accompanying diagram can be used to explain how that braided pattern emerged (Vujakovic 1995, p. 253). Given its importance, it is vital that graphic material in your poster be bold and relevant. Avoid filling up your work with unnecessary pictures.

Creating a simple and comprehensible poster may also require you to convert the written word and numerical information into readily understood graphic form (e.g. bar charts, pie diagrams, and maps). This process is an important element of graphic communication.

When considering the production of figures for a poster, several questions should be addressed. These include:

- can this data be transformed into a figure as opposed to a table? (e.g. a table of population growth figures for Rarotonga since 1900 might be better presented as a line graph);
- what type of figure will best illustrate the point? (e.g. would a cartogram, that is, a form of map in which the sizes of places are adjusted to represent the statistics being mapped, work better than a series of bar charts to illustrate sheep populations in countries of the world?);
- what symbols and colours may be employed to make a greater impression and to better communicate the idea? (e.g. would it be effective to illustrate New Zealand's balance of payments history in a bar graph depicting piles of coins shaded red or black depending on each year's deficit or surplus?) Graphic devices employed may be more readily understood and may therefore be more effective if they incorporate symbols and colours that have evolved through tradition, convention, and public recognition to be representative of their content (e.g. crosses for death, male/female symbols).

Figure 5.3 shows examples of statistics from the same research, which might be incorporated into a poster display in different forms. In general, it is advisable to find a simplified but graphically interesting way of displaying data when producing a poster. Whereas the table might provide the most accurate record of research data collected, it does not enable the viewer to absorb the major trends as quickly as from a graphic depiction. A detailed graph (figure 5.3b), which might be used in a written report, would enable the reader

Figure 5.3: Converting a numerical table into a poster-ready figure

(a)

	COMPANY EXPENDITURE		
Year	Locally	Overseas	Total
1981	5,067,000	1,010,000	$6,077,000
1982	5,328,000	3,126,000	$8,454,000
1983	6,141,000	2,842,000	$8,983,000
1984	7,002,000	6,989,000	$13,991,000
1985	8,269,000	2,354,000	$10,623,000
1986	2,103,000	2,413,000	$4,516,000
1987	2,025,000	4,201,000	$6,226,000
1988	1,567,000	3,368,000	$4,835,000
1989	4,824,000	423,000	$5,247,000
1990	6,041,000	0	$6,041,000

(b)

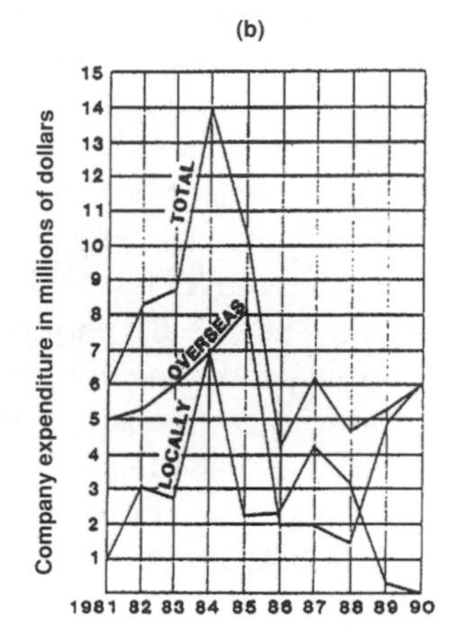

(c)

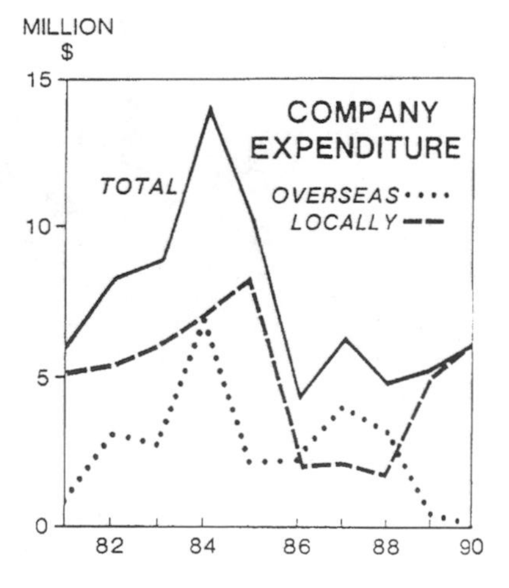

not only to obtain a picture of the relative trends quickly, but also to make a reasonable interpolation of the data. However, a graph such as this is still unnecessarily detailed for a poster display. A simplified and more interesting portrayal (figure 5.3c) enables the viewer to grasp the essential pattern quickly. In this example of a graph suitable for a poster display, interest and clarity is achieved by differentiating linework, type style, and size. The addition of colour might add extra eye appeal.

If you use photographs in your poster, be sure that they are of high quality (e.g. in focus and sharp contrast) and sufficiently large to be interpreted from 1–2 metres away. If the size of the photographed object would not be immediately clear to your audience, provide some idea of scale (Singleton 1984, p. 18). For example, include your camera lens cover in the photograph. Do not forget to provide an appropriate title for the photograph (and for other figures on the poster).

While producing your poster, constantly consider the principles introduced earlier: attention getting, brevity, clarity, directness and careful use of evidence. Providing, of course, that the poster draws from good research, careful application of these principles should contribute to the production of first-class work.

Poster Assessment Schedule

Student Name: Grade: Assessed by:

The following is an itemised rating scale for various aspects of written assignment performance. Sections left blank are not relevant to the attached assignment. Some aspects are more important than others, so there is no formula connecting the scatter of ticks with the final percentage for the assignment. Ticks in either of the two boxes left of centre means that the statement is true to a greater (outer left) or lesser (inner left) extent. The same principle applies to the right-hand boxes. If you have any questions about the individual scales, final comments, final grade or other aspects of this assessment, please see the assessor indicated above.

Quality of argument

Clear statement of question or relationship being investigated					Ambiguous or unclear-statement or purpose
Poster 'stands alone' requiring no additional explanation					Poster is difficult or impossible to comprehend without additional information
Logical and thorough explanation of the quesion/relation being investigated					Illogical or inadequate explanation
All components in presentation given appropriate level of attention					Insufficient treatment of components

Quality of evidence

Argument well supported by evidence and examples					Inadequate supporting evidence or examples
Accurate presentation of evidence and examples					Much evidence incomplete or questionable

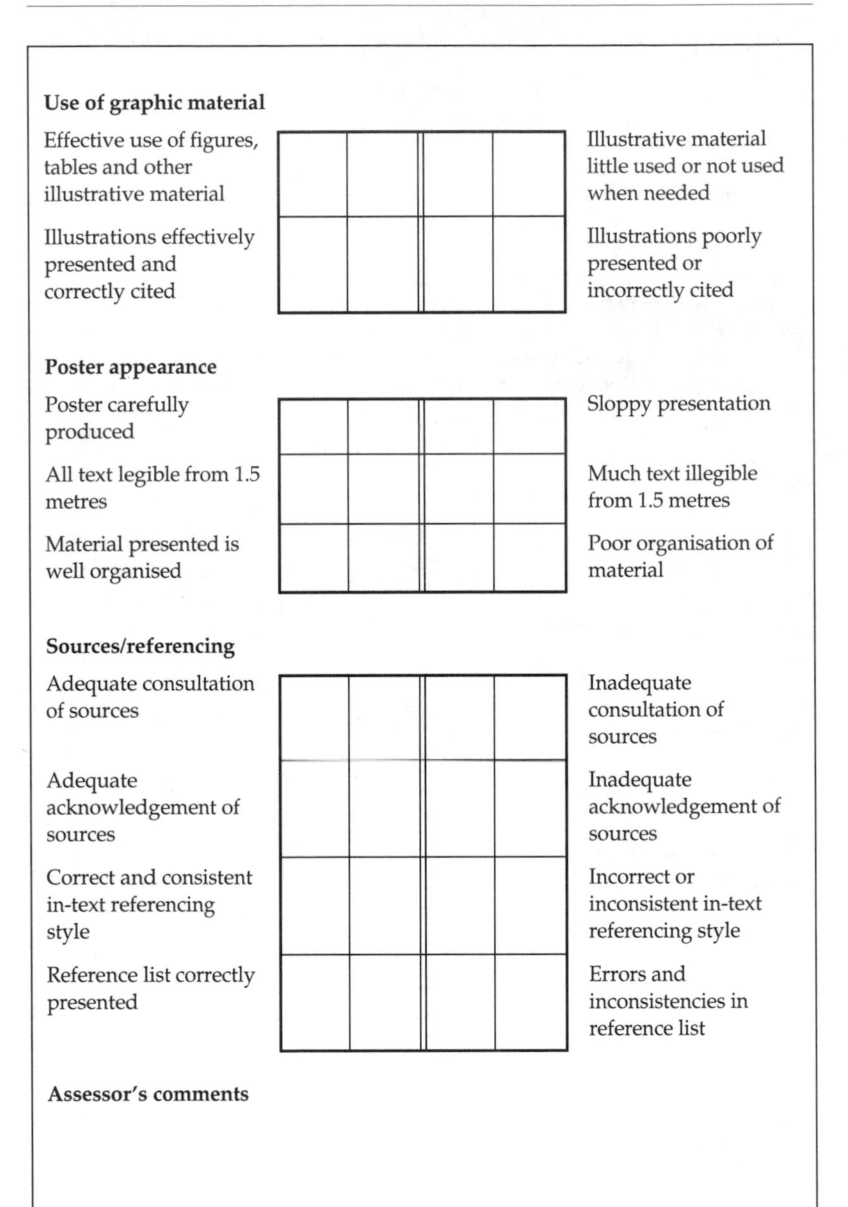

Use of graphic material

Effective use of figures, tables and other illustrative material					Illustrative material little used or not used when needed
Illustrations effectively presented and correctly cited					Illustrations poorly presented or incorrectly cited

Poster appearance

Poster carefully produced					Sloppy presentation
All text legible from 1.5 metres					Much text illegible from 1.5 metres
Material presented is well organised					Poor organisation of material

Sources/referencing

Adequate consultation of sources					Inadequate consultation of sources
Adequate acknowledgement of sources					Inadequate acknowledgement of sources
Correct and consistent in-text referencing style					Incorrect or inconsistent in-text referencing style
Reference list correctly presented					Errors and inconsistencies in reference list

Assessor's comments

6

Preparing and delivering a talk

All the great speakers were bad speakers at first
(Emerson in Mohan, McGregor & Strano 1992, p. 331).

It is not okay to be boring.

Although talking comes naturally to most of us, public speaking remains one of the most frightening things many people can imagine. While it may be difficult to overcome this fear, some understanding of the mechanics of an effective talk, plus a little practice, will help. After a background discussion, this chapter is subdivided into three main parts: the first section deals with the vital matter of preparing for the talk; the second with delivery; and the third with coping with post-talk questions. As with other parts of this book you will find an appropriate assessment schedule at the end of the chapter.

Why Talk?

Believe it or not, lecturers do appreciate the fact that public speaking is one of the greatest fears of most people. While some lecturers themselves have grown accustomed to speaking before a large audience, many still feel some trepidation about speaking to an unknown class, a professional gathering, a community meeting or even a wedding party. They do understand the sleepless nights, sweaty palms, pounding heart, cotton mouth, and jelly-legs which sometimes precede a talk. So, when you are asked by your lecturer to give a prepared talk in class, it is unlikely that the assignment has been set lightly. Lecturers usually have two fundamental objectives in mind when they ask you to give a talk in your geography or environmental studies class.

First and foremost, preparing for and delivering a talk encourages you to organise your ideas, to construct logical arguments, and to otherwise *fulfil the objectives of a university education* (for a discussion, see Jenkins & Pepper 1988, p. 69).

Second, your lecturers also have *vocational objectives* in mind. In many of the jobs in which university-educated geographers, environmental managers, and social scientists find themselves employed (town planning, conservation agencies, professional lobbying), there is an occupationally-related need to make public presentations. Indeed, Patton (1990, p. 428) observes that:

> ...final [written] reports frequently have less impact than the direct, face-to-face interactions I have with primary evaluation users to provide them with feedback about evaluation findings and to share with them the nature of the data. Making oral briefings is...increasingly important...

The ability to communicate has been acknowledged by business and educational leaders to be a central objective of university education. Oral and written communication skills have been identified as having great vocational importance. Indeed, recent studies in Australia and overseas reveal oral and written communication skills to be among the most important abilities a university graduate can have (for examples see Hay 1994a), while a number of international surveys indicate that they are among those most poorly developed by university graduates. One way you might distinguish yourself from other people searching for a job is through your skills in public speaking. The opportunities to demonstrate that skill may become increasingly common. For example, a requirement for many job interviews in the United States, Canada, and Australia is that short-listed candidates give oral presentations to their prospective colleagues and employers.

In short, then, the ability to communicate orally is an important skill sought by employers in Australia and overseas. Graduates who can demonstrate competence in this form of communication are likely to find themselves with an advantage over their less articulate colleagues in the competition for employment.

Although there are other reasons for learning to communicate orally, disciplinary and vocational agendas will usually underpin your lecturers' requests that you contribute to discussion in tutorials, and give impromptu talks, group presentations and more formal individual deliveries. You should regard each of these activities as an opportunity to develop a skill that will make you more effective in your field of study and may also help you to get a job.

Some General Points on Talking

> *Directive for lulling an audience to sleep*
>
> *Wear a dark suit and conventional tie; turn down the lights; close the curtains; display a crowded slide and leave it in place; stand still, read your paper without looking up; read steadily with no marked changes in cadence; show no pictures, use grandiloquent words and long sentences.*
>
> *(Booth 1993, p. 42)*

The following section of this book offers advice on preparing for, presenting, and concluding a successful public presentation. The advice given deals primarily with presenting an extemporaneous speech, the form of oral presentation most commonly used in formal and semi-formal professional settings. Extemporaneous speeches are prepared thoroughly beforehand, but the speaker performs as if talking spontaneously on the subject. Three other types of oral presentation exist: speeches that are read (reserved for very important occasions where slight errors will generate considerable criticism), speeches that are rehearsed and memorised, and impromptu speeches (Windschuttle & Elliott 1994, pp. 342–343).

The guidelines given are perhaps most appropriate for presentations lasting 20 minutes or more, but even for shorter talks the basic principles still apply: plan ahead and know your material.

These guidelines are offered to assist you in the preparation for, and delivery of, your first few public talks; they are not intended to be a prescription for perfection. It is expected that with experience you will develop your own style of presentation, a manner which may be very effective and yet transgress some of the guidelines offered here.

Practice will help you to develop your own style, but you may also want to keep a critical eye on your lecturers and other people giving talks that you attend. Pay close attention to the form and manner of their delivery. Try to pick out the devices, techniques and mannerisms you believe add to or detract from a presentation. Apply what you learn in your own talks.

It is worth remembering, too, that if you have good material and you *care* about conveying your message to your audience, you have already gone a long way towards giving a good talk.

Preparation

You cannot expect to talk competently off-the-cuff on any but the most familiar topics. Effective preparation is critical to any successful presentation. Preparation for a talk should begin some days (at least) in advance of the actual event and certainly not just the night before. Give yourself plenty of time to revise and rehearse. But before you can prepare your talk, several things must first be done.

Establish the context and goals

- **Who is your audience?** Target the presentation to the audience's characteristics, needs and abilities. The ways in which a topic might be developed will be critically influenced by the background and expertise of your listeners (Eisenberg 1992, p. 333). Find out how big the audience will be as this may affect the style of presentation. For example, a large audience will make an interactive presentation somewhat difficult.
- **Where are you speaking?** If possible, visit the venue in which the talk is to be held. Room and layout characteristics can have an effect on the formality of the presentation, the speed of the talk, attentiveness of the audience, and the types of audio-visual aids that can be employed. Check, for example, to see if the talk is to be given in a large room, from a lectern, with a microphone, to an audience seated in rows ...
- **For how long will you speak?** Confirm how much of the time available is for the talk and how much is intended for audience questions. Avoid the embarrassment of being asked to conclude the talk before it is finished or of ending well short of the deadline.
- **Why are you speaking?** The style of presentation may differ depending on your purpose. Are you there to entertain, educate, or persuade?
- **Who else is speaking?** This may influence the audience's reaction to you (Eisenberg 1992, p. 332). It may also require that you take steps to avoid duplication of materials.
- **What is your subject?** Be sure that your subject matches the reason for the presentation. A mismatch may upset, bore, or alienate your audience. A clear sense of purpose will also allow you to focus your talk more clearly.
- **Do your research.** Keeping in mind the purpose of your talk, gather and interpret appropriate and accurate information. Make

a point of collecting anecdotes, cartoons, or up-to-date statistics, which might make your presentation more appealing, colourful, and convincing.

- **Eliminate the dross.** 'The more communication there is, the more difficult it will be for the communication to succeed.' (Goldhaber, in Mohan, McGregor & Strano 1992, p. 340). If you have already written a paper upon which your presentation is to be based, be aware that you will not be able to communicate everything you have written. Carefully select the main points and devote attention to the strategies by which those points can be communicated as clearly and effectively as possible. Courtenay (1992, p. 220) makes the following suggestions.

- List all the things you know or have found out about your subject.
- Eliminate all those items you think the audience might already know about.
- Eliminate anything that is not important for your audience to know. Keep doing this until you are left with one or two new and dynamic points. These should not already be known to your audience and they should be interesting and useful to them. These points should form the basis of your presentation. Similarly, Stettner (1992, p. 226) makes a very good argument for organising a talk around no more than, and no fewer than, three main points.

Organise the material for presentation

Most presentations seem to adopt one of the following five organisational frameworks:

- **chronological**—e.g. the history of geographic thought from the nineteenth century.
- **scale**—e.g. overview of national responses to desertification followed by a detailed examination of responses in a particular area.
- **spatial** —e.g. a description of Japan's trading relations with other countries of the Pacific.
- **causal**—e.g. implications of financial deregulation for the New Zealand insurance market.
- **order of importance**—e.g. ranked list of solutions to the problem of male homelessness in Perth.

- **Ensure that the framework used is appropriate.** Does the organisational framework help make the point of your presentation clear?
- **Give your talk a clear and relevant title.** An audience will be attracted to, and informed by, a good title. Be sure that your title says what the talk is about.

Introduce, discuss, conclude

In most cases, and irrespective of which organisational framework is used, a talk will have an introduction, a discussion, and a conclusion. The introductory and concluding sections of oral presentations are of great importance. About twenty-five per cent of your presentation ought to be devoted to the beginning and end. The remainder of time is spent on the discussion.

Ensure that the framework used is appropriate

- **Make your conceptual framework clear**. This gives the audience a basis for understanding the ideas that follow. In short, let listeners know what you are going to tell them. To do this effectively:

  - **state the topic**—'Today I am going to talk about ...' Do this in a way which will attract the audience's attention.
  - **state the aims or purpose**—Why is this talk being given? Why have you chosen this topic? For what reasons should the audience listen?
  - **outline the scope of the talk**—Let the audience know something about the spatial, temporal, and intellectual boundaries of the presentation. For example, are you discussing Australian attitudes to the environment from a Koori perspective; or offering a geographer's view of British financial services in the 1990s?
  - **provide a plan of the discussion**—Let the audience know the steps through which you will lead them in your presentation and the relationship of each step to the others. It is useful to prepare a written plan for the audience (e.g. on an overhead transparency) which outlines your intended progression.

- **Capture the audience's attention from the outset**. Do this with a rhetorical question, relevant and interesting quotes, amazing facts, an anecdote, startling statements ... Avoid jokes unless you have a real gift for humour.
- **Make the introduction clear and lively**. First impressions are very important.

The Discussion

- **Provide the audience with reasons and evidence to support your views**: Limit discussion to a few main points. Lindsay (1984, p. 48) observes that a rule of broadcasting is that it takes about three minutes to put across each *new* idea. As discussed earlier, do not make the mistake of trying to cover too much material.
- **Present your argument logically, precisely, and in an orderly fashion**: Try producing a small diagram which summarises the main points you wish to discuss. Use this as a basis for constructing your talk.
- **Accompany points of argument with carefully chosen, colourful, and correct examples and analogies**. It is helpful to use examples built upon the experience of the audience at whom they are directed. Analogies and examples clarify unfamiliar ideas and bring your argument to life.
- **Connect the points of your discussion with the overall direction of the talk.** Remind the audience of the trajectory you are following by relating the points you make to the overall framework you outlined in the introduction. For example, 'the third of the three points I have identified as explaining ...'
- **Restate important points.**
- **Personalise the presentation**. This can add authenticity, impact and occasional levity. For example, in discussing problems associated with administering a household questionnaire survey, you might recount an experience of being chased down dark suburban streets by a large, ferocious dog. Avoid overstepping the line between personalising and being self-centred by ensuring that the tales you tell help the audience understand your message.

The Conclusion

- **Cue the conclusion**. Phrases like 'To conclude ...' or 'In summary ...' have a remarkable capacity to stimulate audience attention.
- **Bring ideas to fruition**. Restate the main points in words other than those used earlier in the discussion, develop some conclusions, and review implications.
- **Tie the conclusion neatly together with the introduction.** The introduction noted where the talk is going. The conclusion reminds the audience of the content and dramatically observes the arrival at the foreshadowed destination.
- **Make the conclusion emphatic**. Do not end with a whimper! A good conclusion is very important to an effective presentation,

reinforcing the main idea or motivating the audience (Eisenberg 1992, p. 340).

- **Terminate the presentation clearly**. Saying 'Thank you', for example, makes it clear to the audience that your talk is over. Try to avoid saying things like 'Well, that's the end.'

Prepare text, notes, handouts and visual aids

- **Prepare well in advance.** Mark Twain is reported to have said 'It usually takes more than three weeks to prepare a good impromptu speech' (in Windschuttle & Elliott 1994, p. 341). Twain may have overstated the case a little, but it is fair to consider the talk as the tip of the iceberg and the preparation the much larger submerged section.
- **Prepare a talk, not a speech.** Avoid preparing a full text to be read aloud. A read-aloud presentation is often boring and lifeless. A talk needs to be kept simple and logical. Major points need to be restated.
- **Prepare personal memory prompts.** These might take the form of clearly legible notes, key words, phrases or diagrams to serve as your summary outline of the talk. Put prompts on cards, on the cardboard borders of overhead transparencies, or on note paper, ensuring that each page is numbered sequentially.
- **Revise your script.** Put your talk away overnight or for a few days after you think you have finished writing it. Come back to the script later asking yourself how the talk might be sharpened.
- **Consider preparing a written summary for the audience.** An oral presentation should be designed to present the *essence* of some body of material. You might imagine the talk to be like a trailer for a forthcoming movie. It presents highlights and captures your imagination. If members of the audience want to know more, they should come along to the full screening of the film (i.e. read the full paper). Depending on the circumstances, it may be helpful to prepare for distribution to the audience sufficient numbers of either a full copy of the paper on which a presentation is based or a written summary. In possession of such a document, the audience is better able to keep track of the presentation and you are freer to highlight the central ideas and findings instead of spending valuable time covering explanatory detail. It may be useful to include in your handout copies of tables and other diagrams you will use in the talk.

- **Prepare a limited number of useful visual aids**. Slide projections, models, blackboard sketches, overhead transparencies, video tapes, maps, and charts help to clarify ideas which the audience may have difficulty understanding; hold the audience's attention; and promote interaction with the audience. Do not prepare too many aids as they may defeat these purposes.
- **Make visual aids neat, concise, and simple**. Simple and clearly drawn overhead projections (OHPs) and other illustrations are more easily interpreted and recalled than are complex versions. Sloppily produced visual aids suggest a lack of care, knowledge, and interest.
- **Visual aids ought to be consistent in their style but should not be boring**.
- **Make no more than five or six points on an OHP**. Make each point in as few words as possible (say, about six words per point)
- **Do not include unnecessary text on OHPs**.
- **Produce large and boldly drawn visual aids**. Visuals that can be seen from about 20 metres should be adequate in most cases.
- **Information shown on OHPs should be typed** (or printed neatly).Type and then photocopy onto OHP acetate. The type size of the original document must be sufficiently large (or should be enlarged with a photocopier) to make easily read transparencies. If you write your OHPs, print the text. Do not write cursively.
- **Use upper and lower case text**. This is much easier to read than block capitals.
- **Use line graphs, histograms, and pie charts**. These are usually more effective and easily understood than tables. However, tables can be useful if they are easy to read.
- **Avoid taking graphs or tables directly from a written paper**. These often contain more information and detail than can be comprehended readily. Redraw graphs and redesign tables to make the small number of points you wish to convey.
- **Use a limited range of colours on OHPs.** Up to three colours should be employed. Remember that some colours may evoke certain feelings which add to or detract from the case being argued.
- **Use dark colours on OHPs**. Avoid light colours such as yellow and orange. These cannot be seen clearly.
- **Ensure that all of your OHP will be displayed through the projector**. Leave some space around the margins of each sheet of acetate to avoid the problem of text overlapping the edge of the projector unit.

Rehearse

- **Rehearse before friends or a video-recorder.** Rehearse until there is almost no need to consult prepared notes for guidance. The purpose of rehearsing is to ensure that you have all the points in the right order, not to commit the talk to memory (Dressel 1992, p. 223). In preparing for the talk, it is often useful to make a video recording of a trial presentation. Video cameras are not sensitive to your feelings in the way that an audience of friends and family might be.
- **Time rehearsals.** Match the time available for the talk with the amount of material for presentation. Spoken presentations consume much larger chunks of time than speakers imagine they possibly could. Indeed, most novice speakers are stunned to find out how much longer their presentation takes to deliver than they expected or felt had elapsed while they were talking. Budget for a few extra minutes to compensate for impromptu comments, technical problems, pauses to gather thoughts—or the breath-taking realisation that the audience is not following the tale!
- **Make full use of the visual aids to be used in the talk.** Employing several visual aids can consume time rapidly as you move from one medium, say the overhead projector, to another, say the blackboard. Pay careful attention to time in delivering multi-media presentations.

Final points of preparation

- **Are you dressed for the occasion?** Although the audience's emphasis should be placed on the intellectual merits of your argument, be aware that your style of dress may affect some people's perceptions of the value of your talk. Dress appropriately.
- **Do the visual aids work and how do they work?** Be familiar with the function of any aid that will be used. Do not be so unprepared that you must exasperate your audience with stupid questions like: 'How do I switch this projector on?' You should have checked before your presentation.
- **Can the audience see you and your visual aids?** Before your talk, sit in a few strategically placed chairs around the room to check whether the audience will be able to see you and your visual aids. Consider where you will stand while talking and take care to avoid the problem of your silhouette obstructing the audience's view of OHPs and the blackboard.
- **Is everything else ready?** Are summaries ready for distribution? Are note cards in order ...?

- **Make absolutely clear in your mind the central message you wish to convey.** *This is critical to a good presentation.* If you do not have the message of your talk firmly established in your own mind, you are unlikely to be able to let anyone else know what it is.

Delivery

Remember that people in your audience *want you to do well.* They want to listen to you giving a good talk and they will be supportive and grateful if you are well prepared, even if you do stumble in your presentation or blush and stammer. The guidelines outlined here are a target at which you can aim. No-one expects you to give a flawless presentation.

It will make your presentation more convincing and credible if you remember and act on the fact that the audience comprises *individuals,* each of whom is listening to you. You are not talking to some large, amorphous body. Imagine that you are telling your story to one or two people and not to a larger group. If you can allow yourself to perform this difficult task, you will find that voice inflection, facial expressions and other elements important to an effective delivery will fall into place.

- **Be confident and enthusiastic.** One of the most important keys to a successful presentation is your enthusiasm. You have a well-researched and well-prepared talk to deliver. Most audiences are friendly. All you have to do is tell a small group of interested people what you have to say. Try to instil confidence in your abilities and your message. Do not start by apologising for your presentation. If it is so bad, why are you giving it?
- **Talk naturally, using simple language and short sentences.** Try to relax, but be aware that the presentation is not a conversation in a public bar. Some degree of formality is expected.
- **Speak clearly.** Try not to mumble and hesitate. This may suggest to the audience that you do not know your material thoroughly. Clarity of speech may require that you slow your normal rapid delivery.
- **Project your voice.** Be sure that the most distant member of the audience can hear you clearly.
- **Engage your audience.** Vary your volume, tone of voice and pace of presentation. Involve the audience through use of the word 'you' e.g.—'You may wonder why we used ...' (Lindsay 1984, p. 55)
- **Make eye contact with your audience.** Although this may be rather intimidating, eye contact is very important. As they used to say in the movies, be sure you can see the whites of their eyes!

- **Face the entire audience.** Do not talk to walls, windows, floor, ceiling, blackboard or projector screen. It is the audience—the entire audience—with which you are concerned.
- **Pay attention to audience reaction.** If the audience does not seem to understand what you are saying, rephrase your point or clarify it with an example.
- **Direct your attention to the less attentive members of the audience.** Take care not to focus your presentation on those whose attention you already have.
- **Write key words and unusual words on the blackboard.**
- **Avoid writing or drawing on blackboards or overhead acetate for more than a few seconds at a time.** Long periods devoted to the production of diagrams may destroy any rapport you have developed with your audience.
- **Stop talking when a diagram/slide/map is first shown.** This is to allow the audience time to study the display.
- **Do not stand in front of completed diagrams.**
- **Be sure that overhead and other light projections are sufficiently high for all the audience to see.** As a rule of thumb, make sure the projection is screened higher than the heads of people in the front row of your audience.
- **When you have finished with an illustration, remove it.** The audience's attention will be directed back at you (where it belongs) and will not be distracted.
- **Switch off overhead projectors and other noisy machines when they are not being used.** If this is impossible, it may be necessary to speak more loudly than usual in order to compensate for the whirring of electric cooling fans.
- **Keep to your time limit.** Audiences do not like being delayed. Maintain an even pace throughout the presentation and do not rush at the end. Last minute haste may leave the audience with a poor impression of your talk.

Coping With Questions

The post-presentation discussion which typically follows a talk allows the audience to ask questions and to offer points of criticism. During this time, a number of issues might be raised for discussion and comment (e.g. suggestions for better ways of obtaining and using data; critical comments on or praise for your approach to the topic; questions about bias).

- **Let the audience know whether you will accept questions in the course of the presentation or after the talk is completed.** Be aware that questions addressed during a presentation may disturb the flow of the talk, may upset any rapport developed with the audience, and may anticipate points addressed at some later stage within the presentation.
- **Stay at or near the rostrum throughout the question period.** Question-time is still a formal part of the presentation. Act accordingly.
- **Be in control of the question and answer period.** However, if there is a chairperson, moderation of question-time is their responsibility.
- **Address the entire audience, not just the person who asked the question.**
- **Recognise questions in order.** Take care to receive inquiries from everyone before returning to any member of the audience who has a second question.
- **Search the whole audience for questions.** Compensate for blind spots caused by building piles, the rostrum, and other barriers.
- **Always be succinct and polite in replies.** For two reasons courtesy should be extended even to those who appear to be attacking rather than honestly questioning. First, if the intent of the question has been misinterpreted—with an affront being seen where none was intended—embarrassment is avoided. Second, one of the best ways of defusing inappropriate criticism is through politeness. If, however, there is no doubt that someone is being hostile, keep your cool and, if possible, move closer to the critic. This reduction of distance is a powerful way of subduing argumentative members of an audience.
- **Repeat aloud those questions which are difficult to hear.** This ensures that you heard the question correctly. Repetition is also for the sake of the audience who may not have heard the question either.
- **Clarify the meaning of any questions you do not understand.**
- **Don't conclude an answer by asking the questioner if their query has been dealt with satisfactorily.** Argumentative questioners may take this opportunity to steal the limelight, thereby limiting the discussion time available to other members of the audience.
- **Deal with particularly complex questions or those requiring an unusually long answer after the presentation.** If possible, provide a brief answer when the question is first raised.
- **If you do not know the answer to a question, say so.** Do not try to bluff your way through a problem, as any errors and inaccuracies may call the content of the rest of the talk into question.

- **Difficult questions may be answered by making use of the abilities of the audience.** For example, an inquiry might require more knowledge in a particular field than you possess. Rather than admitting defeat, it is sometimes possible to seek out the known expertise of a specific member of the audience. This avoids personal embarrassment, ensures that the question is answered, and may endear you to that member of the audience whose advice was sought. It also lets other members of the audience know of additional expertise in the area.
- **Smile.** It is over!

Assessment Schedule For a Talk

Student Name: Grade: Assessed by:

The following is an itemised rating scale of various aspects of a formal talk. Sections left blank are not relevant to the assessed assignment. Some aspects are more important than others, so there is no formula connecting the scatter of ticks with the final percentage for the assignment. A tick in the box left of centre means that the criterion has been met satisfactorily. A cross in the right-hand box indicates that improvement is possible. If you have any questions about the individual scales, final comments, final grade or any other aspects of this assignment, please see the assessor indicated above.

First impressions

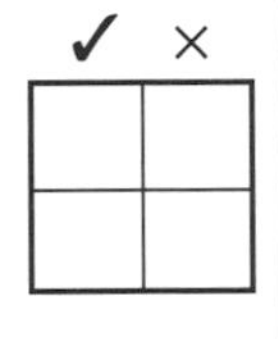

	✓	×
Speaker appeared confident and purposeful before starting to speak		
Speaker attracted audience's attention from the outset		

Presentation structure

Introduction

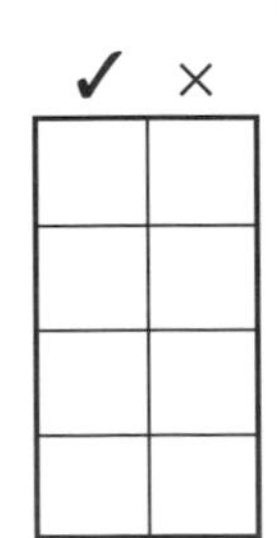

	✓	×
Title/topic made clear		
Purpose of the presentation clear		
Organisational framework made known to audience		
Unusual terms defined adequately		

Presentation structure

Body of presentation

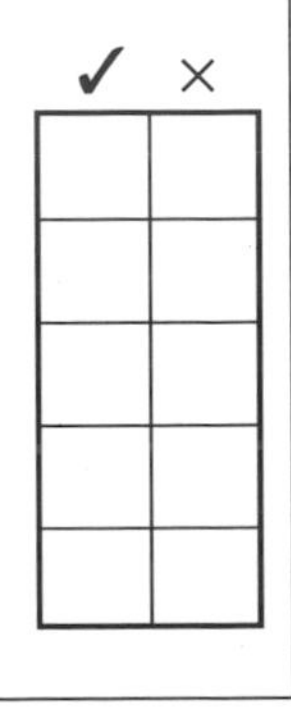

	✓	×
Main points stated clearly		
Sufficient information and detail provided		
Appropriate and adequate use of examples/anecdotes		
Correspondence of presentation content to introductory framework		
Discussion flowed logically		

Presentation structure

Conclusion

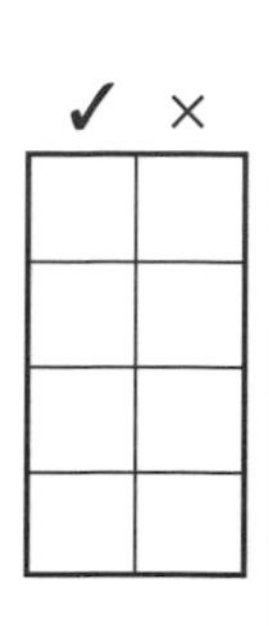

Ending of presentation signalled adequately

Main points summarised adequately/ideas brought to fruition

Conclusion linked to opening

Final message clear and easy to remember

Coping with questions

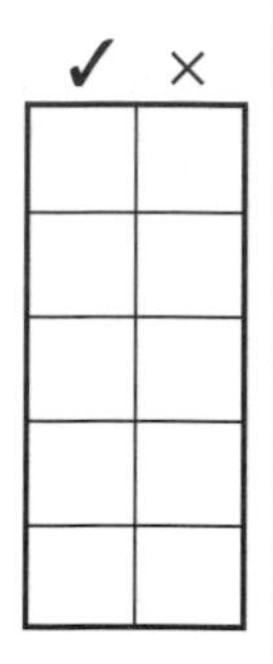

Whole audience searched for questions

Questions addressed in order

Questions handled adeptly

Full audience addressed with answers

Speaker maintained control of discussion

Delivery

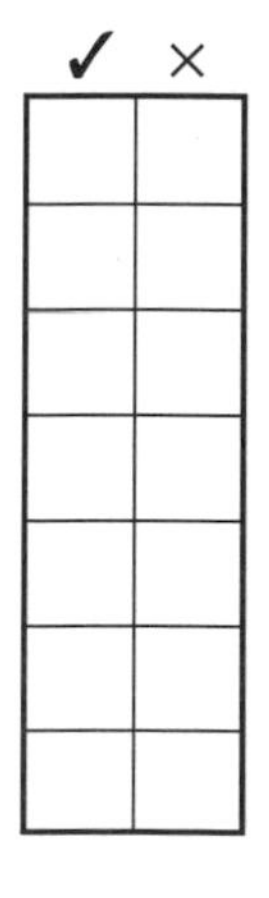

Speech clear and audible to entire audience

Impulsion (engagement and enthusiasm)

Presentation directed to all parts of audience

Eye contact with audience throughout presentation

Speaker kept to time limit

Good use of time without rushing at end

Pace neither too fast nor too slow

Visual aids and handouts—if appropriate

	✓	×
Visual aids clearly visible to entire audience		
Overhead and slide projectors etc. operated correctly		
Speaker familiar with own visual aids (e.g. OHPs, blackboard diagrams)		
Visual aids well prepared		
Effective use made of handouts and/or visual aids		
Handouts well prepared and useful		

General comments: was this an effective talk?

7

Coping with examinations

Some people flourish in examinations, completing their best work under conditions that others might find highly stressful. In part, good performance may be a consequence of a person's particular response to stress, but it is more likely the result of good exam technique. This chapter discusses examinations, their types, and strategies for test success.

Why are Examinations Set?

Examinations are set for three main educational reasons. These are:

- to test your level of factual knowledge;
- to test your ability to synthesise material learned throughout a teaching session; and
- to explore your informed opinion on some specific topic.

These reasons will give you an appreciation of the sorts of things an examiner is likely to be looking for when marking a test. Some tests may seek to fulfil only one of these objectives (e.g. some short answer and multiple choice tests may only be examining your ability to recall information) while others will be looking for all of them (e.g. essay questions or oral examination of a thesis).

Types of Examination

In geography and environmental sciences, there are three basic forms of examination. These, and their subgroups, are listed below.

Closed book

- multiple choice
- short answer
- essay answer

The closed book model of examination requires that you answer questions on the strength of your wits and ability to recall information. No information other than that provided by the examiner for the purposes of the test is permitted.

Open book

- exam room
- take home

In open book exams you are permitted to consult reference materials such as lecture notes, text books and journals. Sometimes the range of texts you may consult will be limited by your examiner.

Oral examination (*viva voce*)

Oral examinations are used most commonly as a supplement to written examinations or to explore issues emerging from an Honours, Masters, or PhD thesis. They may require you to give a brief presentation and then engage in a critical (in the nicer sense of the word) discussion with examiners about the content of your written work.

Of these three forms of examination, the closed book model is certainly the most common. In consequence, most of the discussion which follows focuses on these examinations. Nevertheless, a good deal of the general advice will apply to the other types. Some specific guidance on other test forms is also provided as appropriate.

An important component of success in examinations is good exam technique; this can be broken into two parts:

i preparation
ii sitting the exam.

The discussion that follows reflects this division.

General Advice on Preparing for an Examination

> Review throughout the teaching session

This is perhaps one of the most difficult things for most people to do in preparation for an examination. It is also very important. Try to review course material as the teaching session progresses, beginning with the very first day of classes. You might review by rewriting lecture notes or keeping up-to-date with notes from assigned readings. Not only will this help you remember material when the time comes to sit the exam, but it will also make it easier to understand lecture material as it is presented throughout the teaching session. That can be a major benefit when it is time to complete the examination. Chapters 1 and 2 of Orr's (1984) book offer detailed guidance for organising in-term revision.

Find out about the exam

If past exam papers are available to you, look at them to gain a sense of the likely format of the exam you will sit and the main topics which might be covered. Be aware, however, that the style and content of exams may change from year to year.

Listen for cues from your lecturer about content. Sometimes lecturers will make thinly disguised hints about the content of an exam. Ask teaching staff to let you know what you can expect in the exam in terms of the types of question you may be asked, the time allowed, the materials needed .... Ask, too, if you will be examined on material *not* covered in lectures and tutorials (Committee on Scientific Writing, RMIT 1993, p. 95). Seek some direction about how you might focus your supplementary reading.

Friedman and Steinberg (1989, p. 175) suggest that you should try to anticipate questions and areas which might be in the exam. Although this can be helpful, it can also be a dangerous game for the inexperienced. With Barass (1984, p. 150) I would argue that it is better to have a good overview of all the material covered in a class than it is to have a more developed understanding of a limited range of areas. Comprehension of all the material means that you should be able to tackle competently any question you encounter. If you have narrowed the scope of your revisions, you are gambling.

Be quite sure that you know exactly *when* and *where* the exam is to be held. This is your responsibility and not that of academic staff.

Find a suitable study space

Arrange some comfortable, quiet, and well-lit place where you can study undisturbed. If possible, make sure it is a place where you can lay out papers and books without risk of them being moved regularly. Try to avoid the dining-room table unless everyone plans to eat from plates balanced on their knees for the weeks you are studying.

> Keep to a study schedule

Once you have found out the dates and times of your exams, draft yourself a study calendar. This calendar should allocate particular days to the revision of particular topics. Table 7.1 shows an example in which the fortunate student has only two exams. The student is performing equally well in each subject. Reflecting this balance, the student has allocated 12 study periods to Environmental Management and 11 to Geography. Some times for relaxation, exercise and other day-to-day activities are also set aside.

Table 7.1: Example of an examination study calendar

Date	*A.M. activity*	*P.M. activity*	*Evening*
20 November	End of classes	Relax, buy groceries etc.	Get notes in order
21 November	Get notes in order	Environmental Management	Environmental Management (play squash)
22 November	Environmental Management	Environmental Management	Environmental Management
23 November	Geography	Geography	Relax (swim)
24 November	Geography	Geography	Geography
25 November	Environmental Management (yoga)	Environmental Management	Environmental Management
26 November	Environmental Management (run)	Environmental Management	Environmental Management
27 November	Environmental Management	**Environmental Management Exam**	Relax
28 November	Geography	Geography	Geography (play squash)
29 November	Geography	Geography (swim)	Geography
30 November	**Geography Exam**	Celebrate!	Continue celebrating

If the exams were weighted differently from one another or if the student was performing better in one topic than another, it would be advisable to devote extra time to specific topics as appropriate.

Be sure to stick to your revision schedule.

Concentrate on understanding, not memorising

In most written examinations, you will be required to demonstrate an understanding of the subject material rather than to regurgitate recalled information. In consequence, revision should focus on comprehension first and fact second. Be sure you understand what the course was about and the relationships between content and overall objectives. To do this, check course syllabuses, lecture notes, essay questions, practical exercises, and textbooks. It may also be useful to speak to your lecturer or tutor about particular difficulties. When you have a grasp of the objectives of the course, you will be in a better position to make sense of the content. Understanding the conceptual framework will also allow you to respond in a more critical and informed manner to exam questions than might be possible on the basis of rote learning.

Seek help if you need it

If there are matters you do not understand, ask your lecturer. If you are experiencing emotional or other difficulties which affect your studying, speak to the lecturer or a university counsellor. You will not be the only person facing such problems. Talk them through.

Vary revision practices

To add depth and to consolidate your understanding of course material, use various means of studying for a topic (Barass 1984, p. 147). Set yourself questions, solve problems, organise material, make notes, prepare simple diagrams that might assist with learning and answering questions, and read notes thoughtfully.

Practise answering exam questions

Have a go at answering past exam questions under exam conditions. It may also be helpful to take your trial answers to your lecturer to confirm that you are on the right track.

Another reason for this sort of practising is that you probably do most of your writing on a computer. In an exam, you are likely to be asked to write manually for three hours. Can you do that? What happens to your fingers? Further, word-processors allow you to move paragraphs around and to make revisions quickly. Paper and pen do not offer that liberty and you have to plan your writing much more carefully. So, aside from the intellectual reasons, these two practical reasons should motivate you to practise exam writing.

Maintain your regular diet, sleep, and exercise patterns

Do not make the mistake of popping caffeine tablets and staying up into the early hours of the morning cramming information into your overtired brain unless you already make a habit of that. You run the risk of falling asleep during the exam. Radical changes to your lifestyle are likely to increase levels of stress and may adversely affect exam performance. If you are in the habit of exercising regularly, keep doing that. Most people find that exercise perks them up, makes learning easier, and enhances exam performance. Exercise a little common sense too. Do not make exam revision time the same time as you shock your body with a conversion from couch potato to trainee marathon runner (see Chapter 5 of Orr 1984 for a discussion of fitness and exam performance). You might also consider telling partners, friends and family that you may be a little more difficult to live with while you are studying.

If the exam is in the morning, it is vital that you consume something to raise blood sugar levels. If you do not think you can eat, try drinking an iced coffee, fruit juice or something similar. Do not face an exam on an empty stomach.

Dress appropriately

For a typical closed book examination in a lecture hall or university gymnasium the key is to dress comfortably. Be sure that you will be warm enough or cool enough to function at your optimum level. If you are involved in an oral examination, present yourself in a way which is both comfortable and which suits the formality of the occasion. As a rule of thumb, think about the type of clothing your examiners are likely to be wearing and dress slightly more formally than them. Feel good about how you look. Your performance may match that feeling.

Pack your bag

Make sure you have your student identification card (in most universities you are required to present your ID in order to sit the exam), pencils/pens, ruler, paper, eraser, watch, lucky charms, and a calculator (if required). Exam booklets and scribbling paper will usually be provided. If the weather is hot, you might also want to pack a drink. This is particularly important for summer exams in some poorly insulated exam venues. Take extra warm clothing if it is possible that the venue will be cold.

Get to the right exam in the right place at the right time

Be sure that you know whether the exam is held in the morning or the afternoon. You should also double-check the location of the exam. Every year, some people turn up in the afternoon for a morning exam or arrive at the wrong place. If you do miss the exam for some reason, see your lecturer *immediately*. You will usually find them in their office for the duration of the exam.

Arrive in good time—not too early, not too late. When you are planning your departure for the exam, allow for the possibility of traffic delays, late buses, and bad weather.

Specific advice on preparing for an open book examination

The open book examination is deceptive. At first, you might think, 'what could be easier than a test into which I can take the answers'? Later, perhaps as late as during the exam, you may come to realise that these exams can be a trap for young players. The key lies in the fact that to deal with open book examination questions as quickly and effectively as possible you must know your subject. The open book simply allows you access to specific examples, references and other material which might support *your* answers to questions. You still have to interpret the question and devise an answer. *You* must produce the intellectual skeleton upon which your answer is constructed and upon that place the flesh of personal knowledge, example, and argument drawn from reference material. If you do not understand the background material from which the questions are drawn, you may not be able to perform as well as you should.

The following may be useful additional advice in preparing for an open book exam:

- Study as you would for a closed book examination.
- Prepare easily understood and accessible notes for ready reference.
- Become familiar with the texts you are planning to use. If appropriate, mark sections of texts so that you can identify them easily. Do not mark books belonging to other people or libraries!

Specific advice on preparing for a take home examination

While a take home exam may appear to offer you the luxury of time, preparation remains a key to success. Do not make the mistake of squandering your time trying to get organised *after* you have been given the exam paper. Not only will you be 'burning daylight' but the preparation is likely to be hurried and inadequate.

To prepare for a take home exam:

- Be sure to have at your disposal appropriate reference material with which you are familiar. In short, you should have read through, taken notes, and highlighted a sufficiently large range of reference materials to allow you to complete the examination satisfactorily.
- Arrange all the reference material in a way that allows you to find specific items quickly (for example, alphabetical by author; under subject headings; by date of publication; or by some other method or combination which you find useful).
- If you will be writing the exam on a computer, be sure that there are sufficient supplies of paper, ink cartridges, or printer ribbons to allow you to print the paper when you finish late at night and all the shops are closed.

A Guide to Technique for Written Examinations

Before an examination almost everyone feels tense and keyed up. However, if you have studied effectively and know about the type of exam you will be sitting, the anxiety you feel will probably help you perform at a higher level than if you were quite blasé about the whole affair. Breathe deeply and stride into the exam room with a sense of purpose. You know your stuff and you know what the course was about. Here is the opportunity to prove it!

Read the instructions carefully before beginning

Check to see how long you have to complete the exam, which questions need to be answered and the mark value of each question. It is wise to *repeat* this process after answering the first question (or several in the case of a multiple choice/short answer exam) to confirm that you are doing things correctly. Lecturers find it most disheartening to mark an exam paper by a capable student who has not followed the instructions. It is even more upsetting to be that student.

Work out a timetable

Calculate the amount of time you should devote to each question. Time allocations can be calculated on the basis of marks per question. For example:

A. 3 hour exam (i.e. 180 minutes)

Question 1	10 marks	18 minutes
Question 2	15 marks	27 minutes
Question 3	25 marks	45 minutes
Question 4	50 marks	90 minutes
Total	100 marks	180 minutes

Such a time budget could be modified usefully by allowing time (say 10 minutes) at the end of the exam to proof-read answers. Indeed, proof-reading time can be most valuable.

B. Two day take home exam (i.e. about 16 hours)

Question 1	12 marks	2 hours
Question 2	38 marks	6 hours
Question 3	50 marks	8 hours
Total	100 marks	16 hours

In Example B, an eight hour working day (e.g. 8 am–noon; 1 pm–5 pm) has been used as the basis to calculate the amount of time available for each question of a take home exam. The student in this example might then add some time in the evening for proof-reading answers.

It is most important that you not only allocate your time carefully but also *adhere to your timetable*. It would be stupid, for example, to spend 40 minutes on Q. 1 in Example A above or to spend all of the first day on Q. 1 in Example B. Time discipline may be difficult, but it is a key to examination success.

Read the questions carefully before beginning

In closed book essay examinations, you will usually be given some preparatory time (commonly 10 minutes) in which to read the questions and to make notes on scrap paper. Use this time effectively. Carefully choose the questions you will answer and think critically about their meaning. Take note of significant words and phrases and underline key words. Jot down ideas that spring to mind as you look over the questions. Use the time as a brainstorming session and record your thoughts immediately. Do not rely on your memory. After several hours of answering an exam paper, you may have completely forgotten the brilliant ideas you had for that last question. Your notes will trigger your memory.

Plan your answers

Do not make the mistake of rushing into your answers like the proverbial bull at a gate. Work out a strategy for approaching your answer to each question. Once the exam has begun, use the ideas you jotted down during reading time as the basis of an essay plan for each answer. Make a rough plan of each answer before you begin writing. This might all be done on separate pages in your examination answer booklet. Not only is a plan likely to give a coherent structure to your answers but, if you do run out of time, the marker *may* refer to the plan to gain some impression of the case you were making. However, take care to distinguish essay plans from final answers in your answer booklet (e.g. put a pen stroke through the plan).

If you are doing a multiple-choice exam, read each question and the range of possible answers very carefully.

Begin with the answers you know best

There is usually no requirement for you to answer questions in any particular order. It is often helpful to tackle the easiest questions first to build up confidence and momentum. Further, if you have the misfortune to run out of time, you will have shown your best work.

Answer the questions asked

The most common mistake people make in exams is that they do not answer the question that was asked, sometimes opting to spew forth a prepared answer on a related topic (Friedman & Steinberg 1989, p. 175; Barass 1984, p. 156). Markers want to know what you think and what you have learned about a *specified* topic. As you will appreciate, the right answer to the wrong question will not get you very far!

Examiners are assessing your level of understanding of particular subjects. Do not try to trick them; do not try using the 'shotgun technique', by which you tell all that you know about the topic irrespective of its relevance to the question; do not try to write lots of pages in the hope that you might fool someone into believing that you know more than you do. Concentrate instead on producing focused, well structured answers. *That* will impress an examiner. Clearly, this advice implies that there is no 'correct' amount of writing associated with particular questions.

Attempt all required questions

It is usually easier to get the first 30–50% of the marks for any written question than it is to get the last 30–50%. In consequence, it is foolish to leave any questions unanswered. In the worst case, make an informed guess.

Grab the marker's attention

People marking written exam papers usually have many scripts to assess. They do not want to see the exam question rephrased as the introduction to an essay answer. They probably do not want to read a long introduction to each answer. Instead, they will want you to capture their attention with clear, concise and coherent answers. Spare the padding. Get to the point.

Emphasise important points

Underscore and highlight key points (Friedman & Steinberg 1989, p. 176). You can do this by underlining words and by using phrases that emphasise important matters (e.g. 'The most important matter is ...', 'A leading cause of ...'). You might also find it useful to use headings in essay answers to draw attention to your progression through an argument. Your examiner will certainly find headings useful. Bullet/numbered lists and correctly labelled diagrams can also be helpful.

Support generalisations

Use examples and other forms of evidence to support the general claims you make (Friedman & Steinberg 1989, p. 177). Exam questions and their answers are often very general in scope. Do realise, however, that your answer will be more compelling and will signal your understanding of course material much more effectively if it is supported with appropriate specific materials drawn from lectures, reading, and your own experience than if it rests solely on bald generalisations.

Write legibly

Examiners hate scripts that are difficult to read. It is very difficult to follow someone's argument if frequent pauses have to be made to decipher hieroglyphics masquerading as the conventional symbols of written English. Please try to write legibly. If you have problems writing in a form that can be easily read, write on alternate lines or print to ensure that the examiner is able to interpret your work. Remember, you are trying to communicate your understanding of material as effectively as you can. Handwriting is one component of that overall process of communication.

If you are doing a take home exam, try to type the paper. Indeed, for some people it is easier to engage in the creative process of writing on a computer than it is to write on paper.

Leave space for additions

Begin answers on new pages so that you can add material if time allows. This is particularly important if you have been prudent enough to leave yourself some time for proof-reading. Often, people find that they recall material about one question while they are answering another. It is useful to have the time and space to add those insights.

Keep calm

If you find that you are beginning to panic or go blank, stop writing, breathe deeply and relax for a minute or two. A few moments spent this way should help to put you back on track. Do not give up in frustration and storm out of the exam room. Why run the risk of working out a way through a problem *after* you have left the exam room but while the exam is still on? Whatever you do, do not cheat! Exams are carefully supervised. Copying and other forms of academic dishonesty in examinations do not go unnoticed.

Proofread completed answers

Allow yourself time to proofread your answers. Check for grammatical errors, spelling mistakes, and unnecessary jargon. You may also find time and opportunity to add important matters you missed in the first attempt at the question.

Specific advice on multiple-choice examinations

Examinations involving a multiple choice component can be very challenging and often require you to complete questions very quickly. It is useful to be aware of some of the peculiarities of this form of examination and some of the techniques which can be employed to get the best possible result aside from being conversant with the tested material.

- Go through the test answering all those questions you can complete easily. If there is a question you find difficult, move on and return to it later if you have time.
- Multiple-choice options usually include a number of completely unrealistic options, sometimes added by the examiner for comic relief. Delete those options which are clearly incorrect. Consider humorous responses for deletion, as they are likely to have been included as distracters. This will help you narrow down your choices.
- Read all the answer options before selecting one.
- Avoid extreme answers. For example, if you were asked to state the correct population of New Zealand in 1996 and the options were: (a) 1.1 million (b) 3.6 million (c) 7.6 million (d) 23 million, you would be well advised to avoid option (d) as it is distinctly larger than the other three options.
- Avoid answers that include absolutes (Northey & Knight 1992, p. 117; Burdess 1991, p. 57). In the world in which we live, 'never', 'always', and 'no-one' are rarely true.
- Avoid answers that incorporate terms with which you are unfamiliar (Northey & Knight 1992, p. 117; Burdess 1991, p. 57). Examiners will sometimes use technical words as sirens luring you into shallow waters.
- Do not be intimidated or led astray by an emerging pattern of answers. If, for example, every answer in the first ten questions of the test appears to have been 'b', that does not mean that the next answer ought to be a 'b'. Nor, of course, does it mean that the next answer is not 'b'!
- If there appears to be no correct answer amongst the options provided, choose the option you judge closest to correct.
- If in real doubt about the correct answer, opt for those with long responses. It is often more difficult for an examiner to express a correct idea in few words than it is to express an incorrect one.
- Unless you have been advised that penalties are imposed for giving incorrect answers, answer every question. If you do not know the answer, make an informed guess.
- Do not 'overanalyse' a question. If you have completed a multiple-choice test and are proof-reading your answers, be cautious about

revising your initial response. If you are hesitant about changing your original answer to a different response, you are advised to leave things alone. You probably got it right the first time. Have faith in yourself.

For the sake of your own well-being and for the mental balance of those also sitting the exam, try to avoid behaving in a way which might distract others (e.g. shaking your leg, tapping a pencil or drumming your fingers on the desk). Depending on how stressed others in the class are, you may not survive the exam!

Specific advice on oral examinations

Formal oral examinations are quite rare and hence most people do not get the opportunity to practise them as they might a written or multiple-choice test. Partly as a result of this, the *viva voce* can be quite intimidating. But if you think about it, the oral examination is simply a formalised extension of the sort of discussion you might have had with colleagues, friends and supervisors about the subject you are studying. As such, it should not be too daunting a prospect. Remember, the examiners are real people too. In some instances, they may be quite nervous about the entire process themselves and particularly about their ability to put you at ease so that a genuine, thoughtful discussion can take place.

In preparing yourself psychologically for a *viva voce*, think about its likely aims. If you do not know the aims of the exam, ask your lecturer/supervisor. Generally, the examiners/discussants will want you to fill in detail which you might not have had the opportunity to include in a written paper or in your thesis. The discussants might also wish to use the *viva voce* as a teaching and learning forum. They may want to encourage you to think about alternative ways in which you might have approached your topic and to discuss those alternatives with them. If you can, try to look forward to a *viva voce* as a potentially rewarding opportunity to explore a subject about which *you* may be the best informed.

As specific advice for completing an oral examination, some of the following points may be useful.

- If you do not understand a question, say so. Ask to have the question rephrased. It may not be you that has the problem; it is quite possible that the question does not make sense.
- Take time to think about your answer. You will certainly not be the first person who cannot answer a question as immediately as a talk-show host.

- To give yourself an extra few moments to think about a question, you may find it helpful to repeat the question aloud—but not every time.
- In most examinations, as in a discussion, you should feel free to challenge the examiners' arguments and logic, but be prepared to give consideration to their views.
- If it appears to you that the examiners have not correctly interpreted something you have said or written, let them know precisely what you meant.

Specific advice on take home examinations

Take home examinations present opportunities and challenges. They provide an opportunity to write very good answers to questions. They also offer the challenge of knowing when to stop. It can be very tempting to spend too much time on the exam, gnawing on it as a dog would a bone. If you have been given 48 hours in which to complete an exam, you do not have to work on the questions that long. Neither are you expected to produce some mammoth tome. Instead, you should produce concise, carefully considered, well-argued answers, supported with examples where appropriate. The time available for the exam is not intended for you to write endlessly. The time is provided for you to think carefully and to focus clearly on the question asked.

8

Referencing and language

One important convention and courtesy of academic communication is citing or acknowledging the work of those people whose ideas and phrases you have borrowed. Although this business of referencing is relatively straightforward, many people encounter difficulties with it. This chapter provides an outline of the form and practice of the two most common forms of referencing. The models followed are those outlined by the Australian Government Publishing Service (1994). The chapter also discusses the serious matters of plagiarism, and sexism and racism in language before concluding with some comments on common punctuation problems.

Why Reference?

Referencing is the practice of acknowledging sources of information. You must refer to sources when you quote, copy, paraphrase, or summarise someone else's opinions, theories or data (CUTL 1995, p. 13).

Referencing serves a number of important purposes. It:

- allows readers to evaluate the comprehensiveness of your research and its (philosophical) roots;
- adds weight to your statements by drawing the reader's attention to previous work conducted by other scholars;
- provides readers with the opportunity to confirm the validity of your work (or to repeat it) by checking the source you have cited.

In order to serve this latter purpose, your references need to be *comprehensive*, providing all of the essential details a person needs to find the cited source in any library or bookshop. References should also be as *concise* as is compatible with comprehensiveness and they should be presented in a *consistent* form.

In addition to its honourable functions, accurate and complete referencing is a key to avoiding the serious consequences of misusing academic conventions (plagiarism).

As the rationale for referencing should suggest, referencing is not simply some tedious, nit-picking task required by pedantic professors and lecturers. Referencing is a key to good scholarship. It is also a skill you are likely to need in your post-university career.

General Advice on Referencing

Before considering the different ways in which references may be presented, it is worth offering some general advice on three simple steps that should be followed to ensure you are referencing correctly:

i When you read something, carefully record all the bibliographic details. If you make a practice of photocopying instead of note-taking, make sure you have recorded the full bibliographic details of the book or journal on the front of every photocopied section.
ii When taking notes, always record the page number from which the information has come; in all your tutorial papers, essays, reports, and other written work, you will be required to provide those page references. You will probably find it useful to record the page number from which material has been taken in the margin of your notes. In your notes be sure to distinguish between your own words and direct quotes.
iii When you write your essay, paper, or report, always indicate the source of information or an idea by correct use of an acceptable referencing system.

A Comparison of Two Referencing Systems

There are two basic systems of referencing in geography and environmental sciences: the author-date system (sometimes called Harvard) and the numerical system (sometimes called the endnote system or note system). Of the two, the former is the most widely used. Both are characterised in table 8.1.

The Author-Date (Harvard) System

The Harvard system comprises two essential components, in-text references and a list of references cited. These are discussed, in turn, below.

In-text references

These notes placed within the text of written work provide the reader with a summary of the bibliographic details of the publication being

Table 8.1: Referencing Systems: Main Differences, Advantages, and Disadvantages

Author/date system	*Numerical system*
Referencing in the text of report	
• author and date in parentheses e.g. '... is the aim of geographers' (Smith 1992) • page number included if needed	• consecutive numbers as superior text or superscript e.g. '... is the aim of geographers' [6]
Reference list at the end of report	
• alphabetical list of references on basis of first author's last name	• numbered list of references on basis of order of appearance in text
Advantages	
• allows author and date to be seen in context within the body of the report • saves turning to a list at the end to find the name of a cited source • provides an alphabetical reference list at the end • means that inserting or deleting references is easy	• prevents the text of the report from being interrupted by wordy references • prevents constant repetition of the same in-text references as only a number needs repeating
Disadvantages	
• creates very long author/date entries if there are multiple authors and sources • creates repetition and disruption to the text when the same source is used often	• creates a non-alphabetical reference list at the end • means turning to reference list to match a numerical reference to its source • creates complications if an extra reference needs inserting later • some word processing packages cannot handle footnoting

Adapted from: CUTL (1995)

cited. This summary may be used as a key to finding the full details of the reference, which are provided in the list of references at the end of the document.

The in-text reference presents summary bibliographic details in the following ways:

- if the reference is to *one page*

  As Delaney (1989, p. 50) has made so clear, a significant challenge confronting geography lecturers lies in providing students with the opportunity to link abstract issues with each student's personal understanding of the world.

- if the reference is to *several following pages*

  More patient care inevitably produces a higher exposure to legal risks and a higher volume of liability suits (Bergen 1969, pp. 506–507).

In order to avoid disruption in the flow of the sentence the citation of author, date, and pages used is generally placed at the end of the sentence although (as in the first example) there are occasions when it is better placed within the sentence.

- if the reference is to a *number of authors*

  Several authors (Brown 1981, p. 9; Johnston 1987, p. 52; Smith 1988, p. 16) agree that perception of the environment is a very personal activity.

  Note that each reference is separated by a semi-colon (;).

- if the reference is to a text written by *two or three authors*

  (Brown & White 1996)

  (Brown, White & Green 1996)

  If the authors' names are incorporated into the text, the ampersand is replaced with 'and'. For example:

  Brown and White (1996) confirm the findings of ...

- if the reference is to a single text written by *more than three people*

  Brown et al. (1995, p. 16) argue that ...

  The abbreviation et al. is short for *et alii* meaning 'and others'.

- if the reference is to a *map*

  Low levels of precipitation are evident through much of central Australia (Division of National Mapping 1980).

- if the reference is to work written by a *committee* or an *organisation*

  CSIRO (1987, p. 41) suggests that soil degradation is of major concern to the agricultural community in Australia.

  Natural disasters may present significant difficulties for residents of New Zealand (Earthquake Commission 1995, p. 12).

Occasionally, a publication will have both individual and organisational authors listed. In such cases, it is common practice to treat the individual as author. The organisation is mentioned when giving full details in the list of references cited.

- if the reference is to *one author quoted in the writings of another*

  It can be strongly argued that urbanisation is a major force of population mobility in the third world (Smith, in Davis 1984, p. 219).

You should avoid such references unless tracing the original source is impossible. You are expected to find the original source yourself to ensure that the information has not been misinterpreted or misquoted by the intermediate author.

- if reference is made to information gained by means of *personal communication*

  Certain aspects of the theory remain the subject of investigation (Lethbridge 1987, pers. comm., 2 May).

It is preferred however that references to personal communications be incorporated more fluidly into the text. For example:

> In a telephone interview I conducted on 12 January 1996, Dr. Chris Jones, Director of City Services, revealed that ...

Personal communications are not usually included in the list of references, but if you have cited a number of personal communications, it may be useful to provide a separate list that gives the reader some indication of the credibility of the people cited (i.e. what is the basis for their authority in the context of your work).

- if the reference is to *electronic information*

The style of reference provided is the same as that for individual, group, organisational, and committee authors as outlined above.

> The National Aids Information Clearinghouse (1994) guidelines give clear advice on ...
>
> David Harvey's (1989) work, *The Condition of Postmodernity* is available at an FTP site ...

List of references

This *alphabetically* ordered (by surname of author) list provides the complete bibliographic details of all sources actually referred to in the text. By convention it does not include those sources you consulted but have not cited, although this might be done if there is a particularly good reason to do so (e.g. in a paper critically reviewing the extent to which academic environmentalists have discussed acid rain, it might be useful to provide a list of all relevant works on the subject, even if they have not been discussed in the text).

The following list of examples of correctly formatted references may be useful when you prepare your own lists. Look carefully at the examples and distinguish between the various kinds of work and how each is organised and punctuated. If you have a type of source which is not discussed below, consult the Australian Government Publishing Service *Style Manual*, or perhaps follow the principles of referencing you discern from the examples below. You will note that a convention of referencing is that the second and subsequent lines of each reference are indented. By convention too, reference lists are single spaced and have a blank line between entries.

Article in a periodical or journal

Gregson, N. & Crewe, L. 1994, Beyond the high street and the mall: car boot fairs and the new geographies of consumption in the 1990s, *Area*, vol. 26, no. 3, pp. 261–267.

Complete book

Gold, J.R., Jenkins, A., Lee, R., Monk, J., Riley, J., Shepherd, I. & Unwin, D. 1991, *Teaching Geography in Higher Education: A Manual of Good Practice*, Basil Blackwell, Oxford.

Book, edition other than first

de Blij, H.J. & Muller, P.O. 1986, *Human Geography: Culture, Society, and Space*, 3rd edn, John Wiley, New York.

Chapter in an edited volume

Goudie, A.S. 1993, 'Land transformation', in R.J. Johnston (ed.) *The Challenge for Geography*, Basil Blackwell, Oxford.

Government publication

Australian Bureau of Statistics 1994, *Building Approvals Australia*, Cat. no. 8731.0, ABS, Canberra.

Department of Environment and Planning, South Australia 1982, *Procedures Manual - South Australian Planning System*, DEP, Adelaide.

Monograph published by an organisation

Lakshmanan, T.R. & Chatterjee, L.R. 1977, *Urbanisation and Environmental Quality*, Association of American Geographers Commission on College Geography Resource Paper No. 77, Washington, DC.

Paper in proceedings

Hay, I. 1993, Writing groups in geography, *Peer Tutoring: Learning by Teaching. Proceedings of the Conference held 19-21 August 1993*, Higher Education Research Office, University of Auckland, pp. 101–122.

Pearce, D. 1993, Circuit tourism in Asia and the Pacific, *Proceedings of the Seventeenth Conference of the New Zealand Geographical Society*, New Zealand Geographical Society, Christchurch, pp. 546–551.

Thesis

Kirby, S. 1994, (Hetero)sexing space. Gay men's experiences and perceptions of everyday spaces, BA(Hons) thesis, Flinders University of South Australia.

Unpublished paper

Anderson, E. 1985, Development and trends in EIA in Australia, paper presented to the International Workshop on Environmental Impact Assessment, Wellington, New Zealand.

Map

Department of Lands, South Australia 1987, *Noarlunga*, 3rd edn, 1:25 000, Topographic series.

Newspaper

Legge, K. 1987, 'Labor to cost the "Keating factor"', *Times on Sunday*, 1 Feb., p. 2.

If the article has no obvious author, full bibliographic details should be provided in both the in-text reference and in the list of references:

Canberra Times 24 Jan. 1987, p. 36.
Financial Review 23 Jan. 1994, p. B7.

Video

Down on the farm. (video recording) 30 June 1994, ABC Television.

Electronic information

Australian Bureau of Statistics. 1994, *CDATA91*, version 2.1 revised, (CD-ROM). ABS Electronic Services, Canberra.

Carroll, L. 1991, *Alice's Adventures in Wonderland*, The Millennium Fulcrum edition 2.7a, (online). Available: FTP: quake.think.com, Directory: pub/text/1991, File: alice-in-wonderland.txt

Briefly, for electronic sources, the reference is treated like many others except that a 'type of medium' (i.e. in what electronic form is this information) statement, which might be online, CD-ROM or disk, is added directly after the name/edition description of the item. Further, there is an 'available' statement outlining the electronic access route or address replaces information on publisher and place of publication. Keep the information concise but be sure you provide sufficient for your reader to gain access to the reference.

Multiple entries by same author

If you have cited two or more works written by the same author they should be listed in chronological order by date of publication. If they were written in the same year, add lower case letters to the year of publication to distinguish one from another in both the reference list and the text (e.g. 1987a, 1987b). The assignment of letters is made alphabetically by initial letters of the reference's title.For example:

Smith, R. 1987a, *A Digest of Water Weeds in South Australia*, Bastian Publishers, Adelaide.

—1987b, Water weeds in South Australia, *Journal of Water Science*, vol. 66, no. 4, pp. 6–18.

—1989, Water weed hazards in New Zealand, *Australian and New Zealand Journal of Water Resources*, vol. 13, no. 3, pp. 135–168.

The Numerical System

As with the Harvard system, this system of referencing provides your reader with two pieces of information. Within the text is a superior or superscript numeral (e.g [3]) which refers the reader to full reference details provided either as a footnote (at the bottom of the page) or as an endnote (at the end of the document). If you refer to the same source, say seven times, there will be seven separate note identifiers within the text which relate to that source. You are also obliged to provide seven endnotes or footnotes. This practice is sometimes simplified by providing full bibliographic details in the first endnote/footnote and abbreviated details in the subsequent notes (AGPS 1994, p. 168). In the past, Latin abbreviations such as ibid. (from *ibidem* meaning 'in the same work'), op. cit. (from *opere citato* meaning 'in the work cited') and loc. cit. (from *loco citato* meaning 'in the place cited') were used as part of that abbreviation process, but that is now discouraged and should not be used at all, except in some works in the Humanities (AGPS 1994, p. 180).

The first endnoted or footnoted reference to a work must provide your reader with all the bibliographic information they might need to find the work. The material required is the same as that outlined in the Harvard system, but by convention the details are presented in a different order. The typical order of presentation is as follows:

- author's initials or given name
- surname
- title
- volume, number, edition details
- publisher (may not be appropriate for journals)
- place of publication (may not be appropriate for journals)
- year of publication
- pages

The following are examples (note that in this system, second and subsequent lines of each reference are typically not indented):

[1] N. Gregson & L. Crewe, 'Beyond the high street and the mall: car boot fairs and the new geographies of consumption in the 1990s', *Area*, vol. 26, no. 3, 1994, pp. 261-267.
[2] A.S. Goudie, 'Land transformation', in *The Challenge for Geography*, ed. R.J. Johnston, Basil Blackwell, Oxford, 1993.

The second and subsequent references to a source do not need to be as comprehensive as the first, but they should leave your reader in no doubt as to the precise identity of the reference (e.g. if there are two books by the same author, you will need to provide sufficient detail for the reader to work out which text you are noting). An example of first and second references is provided below:
[1] N. Gregson & L. Crewe, 'Beyond the high street and the mall: car boot fairs and the new geographies of consumption in the 1990s', *Area*, vol. 26, no. 3, 1994, pp. 261–267.
[2] ...
[3] ...
[4] Gregson & Crewe, p. 263.

Notes and Note Identifiers

Sometimes you may wish to let your reader know more about a matter discussed in your essay or report, but believe that extra information to be peripheral to the central message you are trying to convey. An indication that this supplementary information exists may be provided within the text through the use of note identifiers such as symbols (e.g. *, §, ¶, ‡) or, preferably, superscript numbers (e.g. [5]). If you are using the numerical system of referencing, notes should be incorporated into the sequence of your references and should not stand apart from it. If you are using the Harvard system, the notes will be either at the bottom of each relevant page or at the end of the document (before the list of references) in a separate section headed 'Notes'.

Quotation Quirks

There are a few conventions associated with making direct quotes with which you should become familiar.

In making a direct quotation of less than about 30 words you should incorporate the quote into your own text, indicating the beginning and end of the quote with single quotation marks. The in-text reference or numerical reference is usually placed after the closing quotation marks. For example:

> He described Hispaniola and Tortuga as densely populated and 'completely cultivated like the countryside around Cordoba' (Colon 1976, p. 165).

> He described Hispaniola and Tortuga as densely populated and 'completely cultivated like the countryside around Cordoba'[5].

If you are making a direct quotation of more than 30 words, you should not use quotation marks, but rather indent and use single-spacing as shown in the example here:

> One historian makes the point clear:
>
> > Although it included a wide range of human existence, New York was best known in its extremes, as a city capable of shedding the most brilliant light and casting the deepest shadows. Perhaps no place in the world asked such extremes of love and hate, often in the same person (Spann 1981, p. 426).

A space immediately precedes and follows the quote.

Plagiarism and Academic Dishonesty

Plagiarism (from the Latin word for 'kidnapper' (Mills 1994, p. 263)) is a form of academic dishonesty involving the use of someone else's words or ideas as if they are your own. Plagiarism may occur as a result of deliberate and deceptive misuse of another person's work or as the result of ignorance or inexperience about the correct way to acknowledge other work. Plagiarism can take a number of forms including:

- presenting substantial extracts from books, articles, theses, other published or unpublished works (such as working papers, seminar or conference papers, internal reports, computer software, lecture notes or tapes, numerical calculations and data) and other

students' work, without clearly indicating the origin of those extracts by means of quotation marks and references;
- using very close paraphrasing of sentences or whole paragraphs without due acknowledgment in the form of references to the original work;
- quoting directly from a source and failing to indicate that the material is a direct quote.

Academic dishonesty takes other forms, too, and may include:

- fabricating or falsifying data or the results of laboratory, field, or other work;
- accepting assistance from another person in a piece of assessed individual work, except in accordance with approved study and assessment provisions;
- giving assistance, including providing work to be copied, to a person undertaking a piece of assessed individual work, except in accordance with approved study and assessment provisions;
- submitting the same piece of work for more than one topic unless the lecturers have indicated that this procedure is acceptable for the specific piece of work in question.

Universities regard academic dishonesty as a very serious matter and will usually impose very severe penalties. Many universities do not regard ignorance of what constitutes plagiarism as an excuse. You would be wise to re-read the notes above. If you are in any doubt about any aspect of academic dishonesty, speak to your lecturer.

Sexism and Racism in Language

You may remember from the chapter on essays that through writing we can shape the world in which we live. By using sexist or racist language in our writing and speaking we may unwittingly be contributing to those unacceptable forms of discrimination in society. In consequence, as you prepare an essay, report, or talk, try to avoid sexist and racist terminology and ideas. When you have completed your work read through it, ensuring that your language does not unkindly or unfairly discriminate against people. In doing the exercise you may also learn a little about your own attitudes.

Sexist language

Language may be sexist in a number of ways (AGPS 1994, pp. 121–135; Eichler 1991, pp. 136–137; Miller & Swift 1981):

- *use of false generics*—using words which refer to one sex when both are being discussed or which encompass both women and men when, in fact, reference is made to one sex only, e.g. discussing parents when only mothers are being considered;
- *use of 'man'* in compounds, verbs and idioms, e.g. workman, manhole, manning the ship, man and the environment;
- *poor use of pronouns* e.g. the use of he, she, him, his or her to refer to any unspecified person or thing which may be male or female or neither (such as a ship, hurricane or country);
- *trivialisation*—usually sees women's activities denigrated and often implies that women behave more irrationally and emotionally than men, e.g. women 'bicker' whereas men 'disagree'; 'office girl' compared with 'filing clerk';
- *stereotyping*—characterising men or women in ways which emphasise stereotypical characteristics, e.g. men depicted as unemotional, uncaring, clumsy; women depicted as emotional, passive, and nimble;
- *generalisation*—characterising both men and women on the basis of statements which apply to only women or men, e.g. an author writing satirically about the US South once said: 'Who are these people? What are they like? Do they have any pastimes besides fighting, hunting, drinking and writing novels? Do they really sleep with their sisters and bay at the moon?'. The paragraph might have been repaired as: 'Do the men really sleep with their sisters and bay at the moon? Do the women wear crinolines and stash their whiskey behind the camellias?' (Miller and Swift 1980, p. 48);
- *parallel/nonparallel treatment* in nonparallel/parallel situations—this sometimes takes the form of women having their role defined through their relationship with a man, e.g. 'man and wife', 'Mrs. Smith, wife of famous Formula 1 driver, George Smith.' Less commonly the reverse applies, as Dennis Thatcher discovered in the 1980s. Remember, he is Margaret Thatcher's husband.

Racist language

Racism is the discriminatory treatment of people on the basis of their race, ethnicity or nationality. It has its basis in a dichotomy between an 'in-group' and an 'out-group'. Language may be racist a number of different ways (AGPS 1994, p. 135):

- *in-group as norm; out-group as deviation*—the ethnic status of the in-group is rarely mentioned whereas that of out-group members is.

This often happens in news headlines, e.g. 'Greek takes Queensland political position'; 'Japanese gang threat';

- *in-group as individuals; out-group as group*—people in the in-group are often described in terms which reflect their individuality (e.g. educational status, age) whereas members of the out-group have their identity outlined only in terms of association with that group;
- *in-group portrayed positively; out-group portrayed negatively*, e.g. 'whingeing Pom'; 'Kiwi ingenuity'; 'Aussie battler';
- *in-group uses euphemisms to express actions with regard to out-groups*, e.g. 'detainment' of Asian refugees in Australia when, in fact, they appear to have been imprisoned.
- *out-groups described in stereotypical terms*, e.g. Vietnamese immigrants to Australia depicted as being nimble-fingered and therefore suited to some forms of clothing manufacture; Chinese immigrants viewed as having business acumen;
- *ethnic and racial slurs*—these set the out-group apart from the in-group, e.g. derogatory names, slurs, and adjectives such as 'wog', 'coon', 'convict', 'nip';
- *illustrative language representing particular group*—usually, illustrations tend to depict people as white, middle-class, and of Anglo-Saxon heritage, e.g. 'Mr. John Doe'; 'Miss Jane Citizen'.

Racism and sexism are offensive and divisive. Please avoid language which contributes to those, and other, forms of discrimination.

Some Notes on Punctuation[1]

One of the most important skills you should have by the time you have completed your university degree is the ability to communicate clearly. One of the keys to good written communication is correct punctuation. The following notes are intended to help rectify some common problems of written English. Please take the time to read through and absorb this advice.

Comma

,

- breaks up long sentences—e.g. Now there was only standing room in Second Class, the battered yellow coaches were filled to

1 The following section draws heavily from CUTL (1995).

overflowing, and on the curves I could see people on the roofs of the following carriages;

- shows a pause or natural separation of ideas—e.g. After the recommendations were implemented, further evaluations were conducted;
- brackets or separates information in a sentence—e.g. The most common, and most easily rectified, problems in essay writing emerge from incorrect acknowledgment of sources;
- precedes linking words, such as 'but', 'so', 'hence', and 'whereas—e.g. The aim was to examine sustainability, but the experiment failed;
- separates information in a list—e.g. The equipment included one inflatable boat, one motor vehicle and a helicopter.

Full stop

.

- ends a complete sentence—e.g. Geography has not always had a smooth ride.
- ends an abbreviation where the final letter of the abbreviation is not the last letter of the word—e.g. p. for page, ed. for editor;
- shows that a word (or words) has been omitted (usually from a quotation), using three points of ellipsis (...)—e.g. Smith (1993, p. 61) stated: 'The British housing system was ... a cornerstone of the country's welfare state.'

Semi colon

;

- connects two sentences or main clauses which are closely connected, but are not joined with a linking word—e.g. The initial survey revealed a high interest; results showed that further action is appropriate;
- separates complex or wordy items in a list—e.g. The following factors are critical: the environmental impact statement; the government and union policies; the approval of business and council; and public opinion.

Colon
:

- introduces a list or quotation—e.g. The following factors are critical: precipitation, temperature and population.
  —e.g. According to Openshaw (1989, p. 81): 'The fundamental technical change that is underpinning the development of the new post-industrial society is the transformation of knowledge which can be exchanged, owned, manipulated and traded.'

Quotation marks
' ' or " "

- indicates a shorter quotation as part of a sentence—e.g. For our purposes, militarism can be broadly defined as 'a set of attitudes and social practices which regards war and the preparation for war as a normal and desirable social activity' (Mann 1988, p. 166).
- shows the titles of journal articles etc.—e.g. Harvey's paper 'Between space and time' is an example of a recent contribution to the field.

Apostrophe
'

- indicates contractions in verbs—e.g. I'm, we'll, can't. Note that abbreviations of this sort belong to the informal register and are not usually acceptable in academic writing.
- indicates possession, as follows:

> - place the apostrophe at the end of the owner-word, then add a possessive s
>
> e.g. The researcher's results (i.e. the results of one researcher)
> - if the original word ends with an s, place the apostrophe at the end of the owner word without adding a possessive s
>
> e.g. The researchers' results (i.e. the results of more than one researcher)

- It is important to distinguish between it's (= it is) and its (= belonging to it, whether singular or plural).

Capital letters

- Use minimally, especially in titles and headings. Small words such as 'and', 'in', 'the', and 'by' should not be capitalised.
- Use only for a specific and formally named item (e.g. France).

Punctuating numerals

- Write numbers of ten or less in words, except when followed by units of measurement—e.g. 'nine field sites'; 9 mm.
- Place a thin space between the numeral and the unit of measurement, and do not use full stops with units of measurement

Glossary

abstract: short statement outlining the objectives, methods, results, and central conclusions of a research report or paper. An abstract is limited in its length (usually about 100–250 words) and is designed to be read by people who may not have the time to read the whole report.

account for: explain how something came about and why.

acknowledgement: statement recognising the people and institutions to which an author is indebted for guidance and assistance. May be incorporated into the preface/foreword. (See also *citation*.)

analyse: exploring component parts of some phenomenon in order to understand how the whole thing works. It can also mean to examine closely. (Contrast with *synthesise*.)

annotated bibliography: list, in alphabetical order by each author's surname, of works (books, papers) on a specific topic. Each work is summarised and commented upon.

appendix: supplementary material accompanying the main body of a paper, book or report. Typically placed at the back of the document. Includes supporting evidence that would detract from the main line of argument in the text or would make the body of the text too large and poorly structured.

appraise: analyse and judge the worth or significance or something.

argue/argument: a debate which involves reasoning about all sides of an issue and offering support for one or more cases. Typically, you will be asked to present a case for/against a proposition, presenting reasons and evidence for your position. In an argument you should also indicate opposing points of view and your reasons for rejecting them. An argument may be written or spoken.

assess: conduct an evaluation, investigating the pros and cons or validity of some issue or situation. You are usually expected to reach some conclusion on the basis of your research and discussion (e.g. is some situation under consideration, right or wrong, fair or unfair).

author-date system: system of referring to texts cited. Comprises two parts: (i) in-text references, which provide a summary of the bibliographic details of the publication being cited. This comprises author's surname, year of publication and page references; (ii) alphabetically ordered list of references, which provides complete bibliographic details of all sources referred to in the text. (Compare with *numerical system*.)

bar graph: graph in which plotted values are shown in the form of one or more horizontal or vertical bars (column graph) whose length is proportional to the value(s) portrayed. (Contrast with *histogram.*)

bibliography: complete list of works referred to or found useful in the preparation of a formal communication (e.g. essay, book review, poster, report). Less commonly, a bibliography refers to a book listing works available on a particular subject. (See also *references (cited)* and *annotated bibliography.*)

caption: explanatory material printed under an illustration. May also refer to the title or heading above a map, figure, or photograph.

cartogram: form of map in which the size of places depicted is adjusted to represent the statistics being mapped. For example, if one was to produce a map of the world showing sheep populations, New Zealand and Australia would appear to be very large compared with most other countries. Although the physical sizes of places will be altered in the production of a cartogram, efforts are made to preserve both their locations relative to other places and their shapes.

choropleth map: a crosshatched or shaded map used to display statistical distributions (e.g. rates, frequencies, ratios) on the basis of areal units such as nations, states, and regions.

circle graph: see *pie graph.*

citation: formal, written acknowledgement that you have borrowed the work of another scholar. Whenever you quote verbatim (i.e. recite word for word) the work of another person and when you borrow the idea(s) of such people, you must acknowledge the source of that information using a recognised referencing system.

clincher: that part of a paragraph that concludes the paragraph's argument. (See also *topic sentence* and *supporting sentence.*)

comment: make critical observations about the subject matter.

compare: discuss the similarities/differences between selected phenomena (e.g. ideas, places). Be quite sure that you know what you are meant to be comparing. (Often used in conjunction with *contrast.*)

concept: thought or idea that underpins an area of knowledge. For example, the concept of evolution underpins much of biology. The idea that new communications technologies 'compress' distance is significant in geography.

conceptual framework: the logic that underpins an argument or the way in which material is presented. A way of viewing the world and of arranging observations into a comprehensible whole. May be imagined as an intellectual skeleton upon which flesh in the form of ideas and evidence are suspended.

conclusion: that part of a talk, essay, poster or report in which findings are drawn together and implications are revealed.

consider: reflect on; think about carefully.

continuous data: observations which could have any conceivable value within an observed range. Thus, includes fractional numbers, such as halves and quarters. (Compare with *discrete data.*)

contrast: give a detailed account of differences between selected phenomena. (Often used in conjunction with *compare.*)

corroboration: support or confirmation of an explanation or account through the use of complementary evidence. (Compare with replication.)

criticise: provide some judgement on strengths and weaknesses. Back your case with a discussion of the evidence. Criticising does not necessarily require you to condemn an idea.

critique: see criticise.

data region: that part of a graph within which data is portrayed (usually bounded by the graph's axes).

define: explain the basic points or principles of something to provide a precise meaning. Examples may enhance your definition.

demonstrate: illustrate and explain by use of example.

describe: outline the characteristics of some phenomenon. Usually, a description might be imagined to be a picture painted with words. What does the phenomenon look like? What patterns are evident? How big is it? Shape? ... There is no need to interpret.

discrete data: phenomena which may be quantified in whole numbers only, for example animal and human populations. (Compare with *continuous data.*)

discuss/discussion: critically examine, using argument. Present your point of view and that of others. May be written or spoken.

dot map: map in which spatial distributions are depicted by dots representing each unit of occurrence (e.g. one dot represents one person) or some multiple of those units (e.g. one dot represents 1000 sheep).

edit: revise and rewrite. Sometimes implies that some material needs to be deleted.

endnotes: short note placed at the end of a document and identified by a symbol or numeral in the body of the text. A textual 'aside', endnotes provide a brief elaboration of some point made in the text but whose inclusion there might be inappropriate or disruptive to the flow of text. In the numerical system of referencing, endnotes may also include details of reference material cited in the text.

enumerate: list or specify and describe clearly.

essay plan: preparatory framework outlining the basic structure and argument of an essay.

essay: brief literary composition which states clearly what you think and have learned about a specific topic.

evaluate: appraise the worth of something. What are the strengths and weaknesses and which are dominant? Make a judgement.

evidence: information used to support or refute an argument or statement. In forming an opinion or making an argument at university you may need to abandon some practices that may have been considered satisfactory in the past. For example, it is not acceptable for you to state such things as 'it is widely known that ...' or 'most people would say that ...' since in these statements you have not provided any evidence about who the people are, why they say what they do, how they came to their conclusions and so on. In other words, you need to present material that supports or refutes your claim.

examine: investigate critically. Present in detail and critically discuss the implications.

explain: answer 'how' and 'why' questions. Clarify, using concrete examples.

extrapolate: to estimate the value of some phenomenon beyond the extent of known values. Typically, this is done by extending historically known trends into the future. For example, if house prices in Dunedin had been increasing at an average rate of 5% per year for the last 20 years and the median value of a house at the end of last year was $100 000, you might extrapolate from the trend to suggest that the median Dunedin house value will have risen to $105 000 by the end of this year. (Compare with *interpolate*.)

footnote: short note placed at the bottom of a page and identified by a symbol or numeral in the body of the text. A textual 'aside', footnotes provide a brief elaboration of some point made in the text but whose inclusion there might be inappropriate or disruptive to the flow of text. In the numerical system of referencing, footnotes may also include details of reference material cited in the text.

foreword: message about the main text of a book. Usually disconnected from that text because it is written by a different author or because it does not contribute directly to the textual content. (Distinct from *preface*.) An example of a foreword might be a statement by a prominent politician about the timeliness and value of the published volume in which the foreword appears.

freewriting: sometimes used as a step in the production of an essay. Involves (i) 'stream of consciousness' writing without concern for overall structure and direction, followed by (ii) careful revision.

generalisation: a comprehensive statement about all or most examples of some phenomenon made on the basis of a (limited) number of observations of examples of that phenomenon.

Harvard system: see *author–date system*.

histogram: graph in which plotted values are shown in the form of horizontal or, more commonly, vertical bars whose area is proportional to the value(s) portrayed. Thus, if class intervals depicted in the histogram are of different sizes, the column areas will reflect this. (Contrast with *bar graph*.)

hypothesis: supposition or trial proposition used as a starting point for investigation. Usually begins with the word 'that' eg. My hypothesis is that Mt. Ruapehu's 1995 eruption promoted tomato growth in horticultural regions of New Zealand's North Island.

illustrate: make clear through the use of examples or by use of figures, diagrams, maps, and photographs.

indicate: focus attention on or point out.

interpolate: estimate a value of some phenomenon between, and on the basis of, values which are already known. For example, if you knew that the median price of a house in Darwin was $110 000 in January and $120 000 in December, you might interpolate the June value to have been about $115 000. (Compare with *extrapolate*.)

interpret: make clear, giving your own judgement. Offer an opinion or reason for the character of some phenomenon.

introduction: first section in a piece of formal communication (e.g. poster, talk, essay) in which author/speaker tells the audience what is going to be discussed and why.

isoline map: map showing sets of lines (isolines) connecting points of known, or estimated, equal values. Common examples include topographic maps, which show lines of equal elevation (contours), and weather maps, which commonly show isobars (lines of equal atmospheric pressure).

jargon: most commonly, technical terms used inappropriately or when clearer terms would suffice. Less commonly, words or a mode of language intelligible only to a group of experts in the field.

justify: provide support and evidence for outcomes or conclusions.

key: see *legend*.

legend: also known as a key. A brief interpretive statement making sense of the symbols, patterns, and colours used in a map or diagram.

line graph: a graph in which the values of observed (x,y) phenomena are connected by lines. Used to illustrate change over time or relationships between variables.

literature review: comprehensive summary and interpretation of resources (e.g. publications, reports) and their relationship to a specific area of research.

log-log graph: see *logarithmic graph*.

logarithmic graph: graph with one (semi-log) or two (log-log) logarithmic axes. Key intervals on logarithmic axes are based on exponents of ten.

map: graphic device which shows where something is. Graphic representation of a place.

north point: graphic indicator of direction north on map.

note identifiers: symbol or numeral used in text to refer a reader to a reference or to supplementary information in endnotes or footnotes.

numerical system: system of referring to texts cited. Comprises a numeral in superscript within the text which refers the reader to full bibliographic details of the reference provided as a footnote (at the bottom of the page) or an endnote (at the end of the document). Compare with *author–date system.*

outline: describe the main features, leaving out minor details. Alternatively, an outline can be a brief sketch or written plan.

paragraph: a cohesive, self-contained expression of an idea usually constituting part of a longer written document. Typically comprises three parts: *topic sentence, supporting sentence(s)* and *clincher*.

paraphrase: summarise someone else's words in your own.

pie graph: also known as circle graph. Circular shaped graph in which proportions of some total sum (the whole 'pie') are depicted as 'slices'. The area of each slice is directly proportional to the size of the variable portrayed.

plagiarism: presenting, without proper attribution, someone else's words or ideas as your own.

population pyramid: form of histogram showing the number or percentage of people in different age groups of a population.

poster: piece of stiff card to which textual and graphic materials such as maps, tables, and photos outlining the results of some piece of research are affixed.

precis: brief summary of a piece of writing or a talk.

preface: section at the start of a book or report in which the author states briefly how the book came to be written and its purpose. The preface will also usually include acknowledgements unless they are presented separately elsewhere.

prove: demonstrate truth or falsity by use of evidence.

quotation: verbatim (i.e. word for word) copy of someone else's words.

references (cited): complete list of works referred to or found useful in the preparation of a formal communication (e.g. essay, book review, poster, report). Usually, a list of references includes only those sources actually cited (i.e. formally acknowledged). See also *author-date system* and *numerical system*.

relate: establish and show the connections between one phenomenon and another.

replication: with respect to the conduct of research, an account or explanation of some phenomenon may be given weight by experiments or studies that repeat the initial study and yield similar results. (Compare with *corroboration.*)

representative fraction: a form of scale which expresses the relationship between distances on a map or diagram and distances in reality in the form of a fraction. For example, the representative fraction 1: 10 000 (or 1/10 000) means that any one unit of distance on the map (e.g. 1 mm, 1 inch, 1 metre) represents 10 000 of those same units in reality (e.g. 10 00 mm, 10 000 inches, 10 000 metres).

review: make a summary and examine the subject critically.

RF: see *representative fraction.*

scale: an indication provided on a map or diagram of the relationship between the size of some depicted phenomenon and its size in reality. A scale is used most commonly to provide a statement of the relationship between distances on the ground and distances shown on the map. Three forms of scale can be distinguished: (i) a simple statement such as, 1 cm represents 1 km (ii) a graphic device which illustrates the relationship (iii) a representative fraction (e.g. 1: 15 000). The representative fraction is the most versatile of the forms of scale.

scattergram: graph of point data plotted by their (x,y) co-ordinates.

semi-log graph: see *logarithmic graph.*

state: express fully and clearly.

summarise: present critical points in brief, clear form.

supporting sentence(s): that part of a paragraph in which discussion substantiating the paragraph's claim(s) is presented. (See also t*opic sentence* and *clincher.*)

synthesise: build up separate elements into some comprehensible whole. (Compare with *analyse*).

table: systematically arranged list of facts or numbers, usually set out in rows and columns.

topic sentence: that part of a paragraph in which the main idea is expressed. (See also *supporting sentence* and *clincher.*)

trace: describe the development of a phenomenon from some origin(s).

viva voce: oral examination.

why: reasons for.

References and further reading

1 Writing Essays

Anderson, J. & Poole, M. 1994, *Thesis and Assignment Writing,* 2nd ed., Wiley, Brisbane.
A comprehensive review of essay writing mechanics.

Barrett, H.M. 1982, *One Way to Write Anything,* Barnes and Noble, New York.
Includes a detailed chapter on writing effective paragraphs.

Bate, D. & Sharpe, P. 1990, *Student Writer's Handbook,* Harcourt Brace Jovanovich, Marrickville, NSW.
This book provides a detailed review of the mechanics of essay writing e.g. essay outlines, paragraphs, English expression and punctuation. (Perhaps a bit confused in its layout and not written in the correct language for its audience.)

Becker, H.S. & Richard, S.P. 1986, *Writing for Social Scientists. How to Start and Finish your Thesis, Book or Article,* University of Chicago Press, Chicago.
Comprehensive and entertaining volume. Although written for graduate students and staff, this book contains useful advice for undergraduate students.

Betts, K. & Seitz, A. 1994, *Writing Essays and Research Reports in the Social Sciences,* 2nd ed., South Melbourne, Nelson.
Chapter 3 is a helpful discussion on structuring an argument. It also includes sections on the nature of evidence in essays and ways of writing introductions and conclusions.

Booth, V. 1985, *Communicating in Science Writing and Speaking,* Cambridge University Press, Cambridge.
Includes a very helpful review of pre-writing and writing strategies. Enjoyable reading.

Burdess, N. 1991, *The Handbook of Student Skills for the Social Sciences and Humanities,* Prentice Hall, Sydney.
Chapter 3 of this book is a lengthy and detailed discussion of writing skills. It offers a valuable account of matters such as essay planning, writing style, writing conventions and presentation.

Cadwallader, M.L. & Scarboro, C.A. 1982, 'Teaching writing within a sociology course', *Teaching Sociology,* vol. 9, no. 4, pp. 359–382.
Outlines the intellectual functions of writing and offers some advice on how to write. Of limited use.

Clanchy, J. 1985, Improving Student Writing, *HERDSA News,* vol. 7, no. 3, pp. 3-4, 24.
A short note arguing that literacy is discipline-specific and outlining five ways in which lecturers might improve the quality of student writing.

Clanchy, J. & Ballard, B. 1991, *Essay Writing for Students. A Practical Guide*, 2nd ed., Longman Cheshire, Melbourne.
A best selling book that covers the entire essay writing process, including choosing a topic, taking notes, planning the answer, drafting and redrafting, assessment. Well worth reading.

Courtenay, B. 1992, *The Pitch*, Margaret Gee, McMahon's Point, Sydney.

Delaney, E. 1994, Writing an 'A' paper for Professor Ed Delaney, course handout, Department of Geography, University of Colorado, Colorado Springs, CO.

Fletcher, C. 1990, *Essay Clinic. A Structural Guide to Essay Writing*, MacMillan, South Melbourne.
This short book comprehensively outlines steps in the planning and construction of descriptive (word picture), narrative (story telling), discursive (different viewpoints), expository (explanatory), analytical (dismantle and understand) and argumentative (persuasive) essays.

Friedman, S.F. & Steinberg, S. 1989, *Writing and Thinking in the Social Sciences*, Prentice-Hall, New York.
A valuable reference on all stages of the writing process.

Hay, I. & Delaney, E. 1994, 'Who teaches, learns: writing groups in geographical education', *Journal of Geography in Higher Education*, vol. 18, no. 3, pp. 317–334.

Hodge, D. 1994, Writing a good term paper, course handout, Department of Geography, University of Washington, Seattle.

Jones, J. 1985?, *Writing, Setting and Marking Essays*, Higher Education Research Office, University of Auckland.
A large part of Jones' booklet is devoted to a very helpful outline of ways of tackling an essay on the basis of key words within the topic.

Kay, S. 1989, *Writing Under Pressure: the Quick Writing Process*, Oxford University Press, New York.

Lester, J.D. 1993, *Writing Research Papers*, 7th ed., Harper Collins, New York.
Extensive coverage of the essay writing process moving from finding a topic to writing a proposal, doing library research, writing note cards, and eventually, writing the paper.

Lovell, D.W. & Moore, R.D. 1992, *Essay Writing and Style Guide for Politics and the Social Sciences*, Australasian Political Studies Association.

Marshall, L. & Rowland, F. 1993, *A Guide to Learning Independently*, 2nd ed., Longman Cheshire, Melbourne.
Chapters 11 and 12 on essay writing are useful reading.

Miller, C. & Swift, K. 1981, *The Handbook of Non-sexist Writing for Writers, Editors and Speakers*, Women's Press, London.

Mohan, T., McGregor, H. & Strano, Z. 1992, *Communicating! Theory and Practice*, 3rd ed., Harcourt Brace, Sydney.
An overview of the communication process, with some detailed chapters on writing. Emphasis is given to the character of writing and audience responses, and the mechanics of writing particular styles of document (e.g. reports, essays, memos, faxes).

Moxley, J.M. 1992, *Publish Don't Perish. The Scholar's Guide to Academic Writing and Publishing*, Praeger, Westport, Connecticut.

Mullins, C. 1977, *A Guide to Writing and Publishing in the Social Behavioural Sciences*, Wiley, London.

Northey, M. 1993, *Making Sense. A student's guide to research, writing, and style*, 3rd ed, Oxford, Toronto.
Chapters 1, 2, 9, and 11 provide useful material on a variety of issues relating to essay writing.

Northey, M. & Knight, D.B. 1992, *Making Sense in Geography and Environmental Studies*, Oxford University Press, Toronto.
Chapters 2, 4 and 7 offer succinct advice on style in essay writing, editing and the sensitive use of language.

Papadakis, E.P. 1992, Why and what for (four): the basis for writing a good introduction, in D.F. Beer (ed.) *Writing and Speaking in the Technology Professions. A Practical Guide*, IEEE Publications, New York, pp. 71–72.
Papadakis outlines a supposedly fail-safe strategy for introducing technical papers. Although there are some useful hints, not all elements of the approach would work well for all essays in geography and environmental studies.

Peters, P. 1985. *Strategies for Student Writers. A Guide to Writing Essays, Tutorial Papers, Exam Papers and Reports*, Wiley, Brisbane.
Chapter 1 of this book attempts to answer the question 'why write?' In so doing, Peters outlines the differences in forms of inquiry between humanities, social sciences and natural sciences. The title suggests that the book also provides specific advice on writing in various media. It does not.

Samson, J. & Radloff, A. 1992, *In Writing. A Guide to Writing Effectively at the Tertiary Level*, Paradigm, Curtin University, Bentley, Western Australia.

Schwegeler, R.A. & Shamoon, L.K. 1982, 'The aims and process of the research paper', *College English*, vol. 44, no. 8, pp. 817–824.
The authors distinguish between, and discuss the implications of, the difference between students' views of essays (essays are an opportunity to show how much 'good' information you have collected and presented according to academic conventions) and academics' view of essays (an opportunity to analyse, interpret and express an argument).

South Australian College of Advanced Education, Study Skills Team 1989, *Essay Writing*, 2nd ed., SACAE, Adelaide.

Taylor, G. 1989, *The Student's Writing Guide for the Arts and Social Sciences*, Cambridge University Press, Cambridge.
A detailed and successful review of the essay writing process.

Windschuttle, K. & Elliott, E. 1994, *Writing, Researching Communicating*, 2nd ed., McGraw-Hill, Sydney.

Windschuttle, K. & Windschuttle, E. 1988, *Writing, Researching, Communicating*, McGraw-Hill, Sydney.

2 Writing Research Reports and Laboratory Reports

Baker, M., Robertson, F. & Sloan, J. 1993, *The Role of Immigration in the Australian Higher Education Labour Market*, Australian Government Publishing Service, Canberra.

Baylis, P. 1993, 'Writing skills and the scientific world I: reports', in V. Hoogstad & J. Hughes (eds) *Communication for scientific, technical and medical professionals*, MacMillan, South Melbourne.

Beer, D.F. (ed.) 1992, *Writing and Speaking in the Technology Professions. A Practical Guide*, IEEE Press, New York.
This valuable edited collection comprises over 60 short papers on a wide variety of communication skills, including technical report writing. Other topics include oral presentations, running meetings, writing resumes, preparing illustrations, and writing proposals.

Behrendorff, M. 1995, Practical aspects of producing a level three report for the Department of Mechanical Engineering, Department of Mechanical Engineering, University of Adelaide.

Bell, M. 1995, *Internal Migration in Australia 1986–1991: Overview Report*, Australian Government Publishing Service, Canberra.

Betts, K. & Seitz, A. 1994, *Writing Essays and Research Reports in the Social Sciences*, 2nd. ed., Nelson, Melbourne.

Blicq, R.S. 1987, *Writing Reports to Get Results: Guidelines for the Computer Age*, IEEE Press, New York.
This book of over 200 pages offers a serious, no-nonsense review of different report types and how to go about compiling them. Numerous examples are provided. Do not be misled by the volume's publishers, the Institute of Electrical and Electronics Engineers. This book is useful to students and practitioners in other fields.

Brower, J.E., Zar, J.H. & von Ende, C.N. 1990, *Field and Laboratory Methods for General Ecology*, 3d ed., W.C. Brown, New York.

Booth, V. 1993, *Communicating in Science. Writing a Scientific Paper and Speaking at Scientific Meetings*, 2nd ed., Cambridge University Press, Cambridge.
Chapter 1 of Booth's readable and brief book offers helpful advice on writing a scientific paper. Particular attention is devoted to the mechanics and detail of scientific presentation.

Committee on Scientific Writing, RMIT, 1993, *Manual on Scientific Writing*, TAFE Publications, Collingwood, Victoria.

Cooper, B.M. 1964, *Writing Technical Reports*, Penguin, Harmondsworth, Middlesex.

Coventry, G., Daly, J., Evans, M., Lowy, C., McMahon, M. & Roberts, G. 1993, *The Health/Medical Care Injury Case Study Project*, Australian Government Publishing Service, Canberra.

(CUTL) Centre for University Teaching and Learning and Faculty of Engineering. 1995, *Report Writing Style Guide for Engineering Students*, Faculty of Engineering, University of South Australia, Adelaide.

Dane, F.C. 1990, *Research Methods*, Brooks/Cole, Pacific Grove, California.

Day, R.A. 1989, *How to Write and Publish a Scientific Paper*, 3rd ed., Cambridge University Press, Cambridge.
Several chapters of this book are devoted to writing the various sections of a research report.

Eisenberg, A. 1992, *Effective Technical Communication*, 2nd. ed., McGraw-Hill, New York.

Freeman, T.W. 1971, *The Writing of Geography*, Manchester University Press, Manchester.
Now somewhat dated, this book provides a broad overview of fieldwork and writing in geography. Chapter 3 is more narrowly focused on the written

presentation of research results and, although emphasis is given to thesis writing, much of the advice is applicable to report writing.

Friedman, S.F. & Steinberg, S. 1989, *Writing and Thinking in the Social Sciences*, Prentice-Hall, New York.
Chapter 3 includes useful discussions of the role of writing in the research process and the three components of the rhetorical stance: subject, audience and voice. Take care however for in their consideration of voice, the authors imply that sufficient evidence and emotive language are mutually exclusive. Clearly this is incorrect. There is also a helpful review of the appropriate and inappropriate uses of jargon in technical writing.

Gray, D.E. 1970, *So You Have to Write a Technical Report. Elements of Technical Report Writing*, Information Resources Press, Washington, D.C.
A very helpful book providing basic guidance on report writing. Set out in the order in which reports are typically written rather than the order in which they are read, chapters provide straightforward advice on how to write specific sections of a report.

Harwell, G.C. 1960, *Technical Communication*, MacMillan, New York.

Hodge, D. 1994, Guidelines for Professional Reports. Course handout for GEOG 426, Department of Geography, University of Washington, Seattle.

Kanare, H.M. 1985, *Writing the Laboratory Notebook*, American Chemical Society, Washington, DC.
Chapter 3, 'Organizing and writing the notebook', of this American book offers detailed advice on keeping a comprehensive laboratory notebook. Other parts of the text discuss matters such as the legal and ethical aspects of note-taking, patents and invention protection.

Kane, E. 1991, *Doing Your Own Research*, Marion Boyars, London.

Lindsay, D. 1984, *A Guide to Scientific Writing*, Longman Cheshire, Melbourne.

Marshall, L. & Rowland, F. 1993, *A Guide to Learning Independently*, 2nd ed., Longman Cheshire, Melbourne.
Chapter 13 is a useful introductory overview of report writing.

McEvedy, M.R. & Wyatt, P. 1990, *Presenting an Assignment*, Nelson, Melbourne.

Mohan, T., McGregor, H. & Strano, Z. 1992, *Communicating! Theory and Practice*, 3rd ed., Harcourt Brace, Sydney.
Chapter 9 offers an overview of report types, functions, format and style.

Moxley, J.M. 1992, *Publish, Don't Perish. The Scholar's Guide to Academic Writing and Publishing*, Praeger, Westport, Connecticut.
Contains two chapters that outline different ways of writing reports based on qualitative and quantitative research.

Northey, M. & Knight, D.B. 1992, *Making Sense in Geography and Environmental Studies*, Oxford, Toronto.
This book has a short chapter on writing lab reports.

Sayer, A. 1992, *Method in Social Science. A Realist Approach*, 2nd ed., Routledge, London.
Sayer's very influential book includes a useful chapter entitled 'Problems of explanation and the aims of social science' which introduces differences between 'intensive' (essentially qualitative) and 'extensive' (essentially quantitative) research and their implications for the process of research and the communication of results. However, this chapter is challenging reading.

Sides, C.H. 1992, *How to Write and Present Technical Information*, 2nd. ed., Cambridge University Press, Oakleigh, Victoria.

Walker, J.R.L. 1991, 'A student's guide to practical write-ups', *Biochemical Education*, vol. 19, no. 1, pp. 31–32.

Windschuttle, K. & Elliott, E. 1994, *Writing, Researching, Communicating. Communication Skills for the Information Age*, 2nd ed., McGraw-Hill, Sydney.
This long, comprehensive book has a number of useful, short chapters on writing administrative, management and annual reports.

Woodford, F.P. 1967, 'Sounder thinking through clearer writing', *Science*, vol. 156, no. 3776, p. 744.

3 Writing Reviews, Summaries, and Annotated Bibliographies

Association of College and Research Libraries. 1994, Guidelines for reviewing electronic media, *Choice Current Reviews for Academic Libraries*, American Library Association, Middletown, Connecticut.
These pages of advice provide a succinct outline of issues which might usefully be considered in the professional review of software and databases.

Barrett, H.M. 1982, *One Way to Write Anything*, Barnes and Noble, New York.

Bergman, B.A. 1978, 'Do's and don'ts of book reviewing', in S.E. Kamerman (ed.) *Book Reviewing*, Writer, Boston.

Berry, B.J.L. 1993, 'Canons of reviewing revisited', *Urban Geography*, vol. 15, no. 1, pp. 1–3.
This short note by Brian Berry includes a reprint of very wise 1964 advice to reviewers by Wesley C. Calef.

Bricklin, B. 1978, 'Reviewing for specialized journals', in S.E. Kamerman (ed.) *Book Reviewing*, Writer, Boston.

Burdess, N. 1991, *The Handbook of Student Skills for the Social Sciences and Humanities*, Prentice Hall, New York.
Chapter 3 includes short sections on writing abstracts and book reviews.

Calef, W.C. 1964, Canons of Reviewing, Illinois State University.

Clanchy, J. & Ballard, B. 1991, *Essay Writing for Students. A Practical Guide*, Longman Cheshire, Melbourne.
Chapter 9 includes a succinct description of how to write a review. Appendix 10 also outlines some of the criteria used by staff in the disciplines of political economy and women's studies when assessing book reviews.

Day, R.A. 1989, *How to Write and Publish a Scientific Paper*, Cambridge University Press, Cambridge.

Friedman, S. & Steinberg, S. 1989, *Writing and Thinking in the Social Sciences*, Prentice Hall, Englewood Cliffs, New Jersey.

Gould, P. 1993, *The Slow Plague. A Geography of the AIDS Pandemic*, Blackwell, Cambridge, Massachusetts.

Gulley, H.E. 1995, 'Review of "Trade and Urban Development in Poland" ', *Annals of the Association of American Geographers*, vol. 85, no. 3, pp. 606–607.

Hay, I. 1995a, Non-Sexist Research Methods, Class Notes for GEOG 3009, Flinders University of South Australia, Adelaide.

Hay, I. 1995b, 'Writing book and article reviews', *Journal of Geography in Higher Education*, vol. 19, no. 3.
This paper is an earlier version of the chapter presented in this book.

Kearns, R. 1994, Review of 'New Zealand's Ageing Society: The Implications', *New Zealand Geographer*, vol. 50, no. 2, p. 62.

Kenny, H.A. 1978, 'The basics of book reviewing', in S.E. Kamerman (ed.) *Book Reviewing*, Writer, Boston.

Kirsch, R. 1978, 'The importance of book reviewing', in S.E. Kamerman (ed.) *Book Reviewing*, Writer, Boston.

Ley, D.F. & Bourne, L.S. 1993, 'Introduction: The Social Context and Diversity of Urban Canada' in L.S. Bourne & D.F. Ley (eds) *The Changing Social Geography of Canadian Cities*, McGill-Queen's University Press, Montreal & Kingston.

McEvedy, M.R. & Wyatt, P. 1990, *Presenting an Assignment*, Nelson, Melbourne.

Northey, M. & Knight, D.B. 1992, *Making Sense in Geography and Environmental Studies*, Oxford University Press, Toronto.
See chapter 6, especially, which provides advice on writing summaries/precis, analytic book reports and literary reviews (i.e. focusing on a theme and requiring coverage of several books).

Smith, J. 1995, Review of 'A Continent Transformed: Human Impact on the Natural Vegetation of Australia', *Australian Geographical Studies*, vol. 33, no. 1, pp. 133–134.

South Australian College of Advanced Education 1989, Article reviews and annotated bibliographies, Adelaide.

4 Preparing Maps, Figures, and Tables

Ahrens, D.C. 1991, *Meteorology Today. An Introduction to Weather, Climate and the Environment*, 4th ed., West Publishing, St. Paul.

Australian Bureau of Statistics. 1994a, *Australia Yearbook*, Cat. no. 1301, AGPS, Canberra.
________. 1994b, *Australian Social Trends*, Cat. no. 4102.0, ABS, Canberra.
________. 1994c, *Labour Statistics Australia 1993*, Cat. no. 6101.0, AGPS, Canberra.
________. 1994d, *Monthly Summary of Statistics Queensland, October 1994*, Cat. no. 1304.3, ABS, Brisbane.
________. 1994e, *Monthly Summary of Statistics West Australia, December 1994*, Cat. no. 1305.5, ABS, Perth.
________. 1994f, *South Australian Yearbook*, Cat. no. 1301.4, ABS, Canberra.
________. 1994g, *Statistics—A Powerful Edge*, Cat. no. 1331.0, ABS, Canberra.

Australian Government Publishing Service. 1988, *Style Manual*, 4th ed., Australian Government Publishing Service, Canberra.

Balchin, W.G.V. 1985, 'Graphicacy comes of age', *Teaching Geography*, vol. 11, no. 1, pp. 8–9.

Birch, T.W. 1964, *Maps, Topographical and Statistical*, 2nd ed., Oxford University Press, London.

Campbell, W.G., Ballou, S.V. & Slade, C. 1986, *Form and style. Theses, reports, term papers*, 7th ed., Boston, Houghton Mifflin.

Cartography Specialty Group of the Association of American Geographers. 1995, 'Guidelines for effective visuals at professional meetings', *AAG Newsletter*, vol. 50, no. 7 (July), p. 5.

Coggins, R.S. & Hefford, R.K. 1966, *The Practical Geographer*, 2nd ed., Longman, Camberwell, Victoria.

Committee on Scientific Writing, RMIT 1993, *Manual on scientific writing*, TAFE Publications, RMIT, Collingwood, Victoria.

Day, R.A. 1989, *How to write and publish a scientific paper*, 3rd ed., Cambridge University Press, Oakleigh, Victoria.

Department of Environment and Land Management 1993, *The state of the environment. Report for South Australia 1993*, Department of Environment and Land Management, Adelaide.

Dickinson G.C. 1973, *Statistical Mapping and the Presentation of Statistics* (2nd ed.) Edward Arnold, London.

Ebel, F.E., Bliefert, C. & Russey, W. 1987, *The Art of Scientific Writing from Student Reports to Professional Publications in Chemistry and Related Fields*, VCH Publishers.

Eisenberg, A. 1992, *Effective Technical Communication*, 2nd ed., McGraw-Hill, New York.

Garnier, B.J. 1966, *Practical Work in Geography*, Edward Arnold, London.

Gerber, R. 1990–1, 'Designing graphics for effective learning', *Geographical Education*, pp. 27–33.

Gerber, R.V. 1977, 'Audio-visuals in Geography', *Geographical Education*, vol. 3, pp. 25–42.

Green, C.C. & Lowing, G.D. 1972, *Practical Geography. A Workbook for Senior Students*, L/S Publishing, Victoria, Australia.

Hodgkiss A.G. 1970, *Maps for Books and Theses*, David and Charles, London.

International Cartographic Association 1984, *Basic Cartography for Students and Technicians*, vol. 1, International Cartographic Association, Great Britain.

Jacaranda Atlas Programme. 1984, *Skills Book for Secondary Schools*, Jacaranda Press, Queensland.

Jennings, J.T. 1990, *Guidelines for the Preparation of Written Work*, 4th ed., University of Adelaide Roseworthy Campus.

Jones, F.G. & Ollier, C.D. 1983, *Basic Geography Skills*, MacMillan, South Melbourne.

Krohn, J. 1991, 'Why are graphs so central in science', *Biology and Philosophy*, vol. 6, no. 2, pp. 181–203.
This paper critically questions the prominence, use and significance of graphics in science.

Linacre, E. & Hobbs, J. 1977, *The Australian Climatic Environment*, John Wiley, Brisbane.

Mohan, T., McGregor, H. & Strano, Z. 1992, *Communicating! Theory and Practice*, Harcourt Brace, Sydney.

Monkhouse, F.J. & Wilkinson, H.R. 1971, *Maps and Diagrams*, 3rd ed., Methuen, London.

Monmonier, M. 1991, *How to Lie with Maps*, University of Chicago Press, Chicago.

Moorhouse, C.E. 1974, *Visual Messages*, Pitman, Carlton, Victoria.
Chapter 7 is a very handy outline of graphics. Well worth consulting.

New Zealand Tourism Board 1995, *New Zealand Where to Stay Guide*, New Zealand Tourism Board, Wellington.

Peters, P. 1985, *Strategies for Student Writers. A Guide to Writing Essays, Tutorial Papers, Exam Papers and Reports*, Brisbane, Wiley.

Rowntree, D. 1990, *Teaching through self-instruction*, Kogan Page, London.

Schmid, C.F. 1983, *Statistical Graphics. Design Principles and Practices*, John Wiley, New York.

Scott, L. & Laws, K. 1983, *Mapping and Statistical Skills for Secondary Students*, Jacaranda Press, Queensland.

Scott, L. 1984, *Our World in Change. People and Resources*, Jacaranda Press, Queensland.

Sides, C.H. 1992, *How to Write and Present Technical Information*, 2nd ed., Cambridge University Press, Oakleigh, Victoria.

Sullivan, M.E. 1993, 'Choropleth mapping in secondary geography: an application for the study of middle America', *Journal of Geography*, vol. 92, no. 2, pp. 69-74.

Szoka, K. 1992, A guide to choosing the right chart type, in D.F. Beer (ed.) *Writing and Speaking in the Technology Professions. A practical guide*, IEEE Publications, New York, pp. 44–47.

Toyne, P. & Newby, P.T. 1971, *Techniques in Human Geography*, MacMillan, London.
This text is now rather old, but chapter 3 remains one of the most succinct and comprehensive discussions available to students on the visual representation of data.

University of Adelaide, Department of Geography 1991, *Geography 1B Society and the Physical Environment—Practical Handbook*, Adelaide.

Wainer, H. 1984, 'How to display data badly', *The American Statistician*, vol. 38, no. 2, pp. 137–147.
A fascinating review of twelve techniques for displaying data badly! Well worth reading.

Windschuttle, K. & Windschuttle, E. 1988, *Writing, Researching, Communicating*, McGraw-Hill, Sydney.

World Almanac 1995, Newspaper Enterprise Association, New York.

5 Preparing Posters

Committee on Scientific Writing RMIT 1993, *Manual on Scientific Writing*, TAFE Publications, Collingwood, Victoria.

Day, R.A. 1989, *How to Write and Publish a Scientific Paper*, Cambridge University Press, Cambridge.

Hay, I. & Miller, R. 1992, 'Application of a poster exercise in an advanced undergraduate geography course', *Journal of Geography in Higher Education*, vol. 16, no. 2, pp. 199–215.
Essentially intended to be read by lecturers, this paper critically discusses a strategy for incorporating a poster exercise into an upper-level class. Special emphasis is given to the rationale for conducting the project and to the practical dimensions of poster production, much of which is elaborated upon in this book.

Howenstine, E., Hay, I., Delaney, E., Bell, J., Norris, F., Whelan, A, Pirani, M., Chow, T. & Ross, A. 1988, 'Using a poster exercise in an introductory geography course', *Journal of Geography in Higher Education*, vol. 12, no. 2, pp. 139–147.
Written by a group of Teaching Assistants, this paper discusses the purpose, design, implementation, and value of a poster exercise in the instruction and promotion of geography at the University of Washington (Seattle) and elsewhere.

Jenkins, A. 1994, 'Conveying your message by a poster', in D. Saunders (ed.) *The Complete Student Handbook*, Blackwell, Oxford.
An easily-read chapter which is written at a more general level than the one in this book. By comparison, Jenkins' discussion has less detail on preparing posters and a little more on techniques for presenting posters within a poster session (i.e. with other posters).

Larsgaard, M. 1978, *Map Librarianship*, Libraries Unlimited, Littleton, Colorado.
Includes discussion of some useful design principles.

Lethbridge, R., 1991, *Techniques for Successful Seminars and Poster Presentations*, Longman Cheshire, Melbourne.
This short book briefly introduces poster design principles and goes on to provide a comprehensive review of technical aspects of preparing illustrations. Indeed, the book's title may be a little misleading given the volume's overall emphasis on graphic production techniques.

Mills, V. 1967, *Making Posters*, Studio Vista, London.
This book approaches poster construction from a graphics arts perspective, but it does provide some useful information on colour, lettering and design for those people preparing academic posters.

Sim, R. 1981, *Lettering for Signs, Projects, Posters, Displays*, 2nd ed., Learning Publications, Balgowlah, Australia.

Simmonds, D. & Reynolds, L. 1989, *Computer Presentation of Data in Science*, Kluwer Academic, Dordrecht.
Includes a very short section on poster layout.

Singleton, A. 1984, *Poster Sessions. A Guide to their Use at Meetings and Conferences for Presenters and Organisers*, Elsevier, Oxford.
A good, succinct discussion of posters, poster sessions and their uses.

Vujakovic, P. 1995, 'Making posters', *Journal of Geography in Higher Education*, vol. 19, no. 2, pp. 251–256.
Readable and concise discussion of poster production.

6 Preparing and Delivering a Talk

Barnacoat, M. 1993, 'Presentation skills: the psycho-social approach', in V. Hoogstad & J. Hughes (eds) *Communication for Scientific, Technical and Medical Professionals*, MacMillan, South Melbourne.
A helpful chapter which discusses presentation strategies and ways of overcoming speaker fear.

Beer, D.F. (ed.) 1992, *Writing and Speaking in the Technology Professions. A Practical Guide*, IEEE Publications, New York.
Part 7 of this edited collection contains a number of good, short papers on speaking effectively to groups.

Booth, V. 1993, *Communicating in Science. Writing a Scientific Paper and Speaking at Scientific Meetings*, 2nd ed., Cambridge University Press, Cambridge.
Chapter 2 is an entertaining and informative exhortation to speak well at professional gatherings.

Bryant, D.C. & Wallace, K.R. 1962, *Oral Communication*, 3rd ed., Appleton-Century-Crofts, New York.

Burdess, N. 1991, *The Handbook of Student Skills for the Social Sciences and Humanities*, Prentice Hall, New York.
See chapter 1 for a helpful review of seminar presentations. Burdess takes the innovative step of discussing the role of the audience in an oral presentation.

Calnan, J. & Barabas, A. 1972, *Speaking at Medical Meetings. A Practical Guide*. Heinemann, London.
A marvellous little book. Although some of the material applies only to specific fields within the medical profession, most of the content of this easily-read, humorous volume is applicable to speakers in other domains.

Campbell, J. 1990, *Speak for Yourself. A Practical Guide to Giving Successful Presentations, Speeches and Talks*, BBC Books, London.
At about 150 pages, this is the most comprehensive review on preparing and presenting a talk I have found. The book accompanies a BBC video of the same title.

Carle, G. 1992, 'Handling a hostile audience with your eyes', in D.F. Beer (ed.) *Writing and Speaking in the Technology Professions. A Practical Guide*, IEEE Publications, New York, pp. 229–231.
Carle's paper outlines an interesting three-step strategy for dealing with hostile audiences.

Clerehan, R. 1991, *Study Skills Handbook for Tertiary Students*, Language and Learning Services, Monash University, Melbourne.

Committee on Scientific Writing RMIT. 1993, *Manual on Scientific Writing*, TAFE Publications, Collingwood, Victoria.
Chapter 4 includes a helpful and short discussion on a talk-planning strategy called clustering.

Courtenay, B. 1992, *The Pitch*, Margaret Gee, McMahon's Point.
This book comprises a series of short articles written by Courtenay—acclaimed author of *The Power of One*, *Tandia*, and *April Fool's Day*—for the *Australian* newspaper. A number of the articles discuss the business of making an oral presentation.

Daniel, P.A. 1991, 'Assessing student-led seminars through a process of negotiation', *Journal of Geography in Higher Education*, vol. 15, no. 1, pp. 57–62.
Includes an assessment schedule for student seminar (interactive) presentations.

Day, R.A. 1989, *How to Write and Publish a Scientific Paper*, 3rd ed., Cambridge University Press, Cambridge.
This is not a particularly helpful book, offering only a cursory look at talking.

Dressell, A. 1992, 'Authenticity beats eloquence', in D.F. Beer (ed.) *Writing and Speaking in the Technology Professions. A Practical Guide*, IEEE Publications, New York, pp. 223–224.

Dudley, H. 1977, *The Presentation of Original Work in Medicine and Biology*, Churchill Livingstone, Edinburgh.

See chapter 5 for a discussion on giving talks and a helpful section on coping with questions.

Eisenberg, A. 1992, *Effective Technical Communication*, 2nd ed., McGraw-Hill, New York.
Chapter 15 is a lengthy discussion on preparing and giving a talk. Some useful notes on aspects of body language and on writing out a talk are included.

Grimmond, T. 1995, *How to Deliver Effective Presentations*, Adelaide.
Terry Grimmond of Flinders University travels Australia giving talks on giving talks. His entertaining and informative sessions have this companion text, which is both valuable and comprehensive.

Harwell, G.C. 1960, *Technical Communication*, MacMillan, New York.
Chapter 9 covers oral communication and includes a discussion of body language.

Hay, I., 1994a, 'Justifying and applying oral presentations in geographical education', *Journal of Geography in Higher Education*, vol. 18, no. 1, pp. 43–55.

Hay, I. 1994b, 'Notes of guidance for prospective speakers', *Journal of Geography in Higher Education*, vol. 18, no. 1, pp. 57-65.

Hay (1994a) and (1994b) are earlier versions of the chapter presented in this book.

Hay, I. & Miller, R. 1992, 'Application of a poster exercise in advanced undergraduate geography classes', *Journal of Geography in Higher Education*, vol. 16, no. 2, pp. 199–215.

Hughes, J. 1993, 'Oral communication: preparation, planning and performance', in V. Hoogstad & J. Hughes (eds) *Communication for Scientific, Technical and Medical Professionals*, MacMillan, South Melbourne.
A pithy chapter discussing a variety of matters including speech organisation and delivery (including dress, body language, grammar).

Jenkins, A. & Pepper, D. 1988, 'Enhancing students' employability and self-expression: how to teach oral and groupwork skills in geography', *Journal of Geography in Higher Education*, vol. 12, no. 1, pp. 67–83.
Includes an extensive discussion of the academic and vocational rationale for teaching and learning oral communication skills.

Jones, M. & Parker, D. 1989, 'Research report development of student verbal skills: the use of the student-led seminar', *The Vocational Aspect of Education*, vol. 41, no. 108, pp. 15–19.
In its introduction this paper outlines some of the vocational reasons for learning to speak effectively.

Kenny, P. 1982, *A Handbook of Public Speaking for Scientists and Engineers*, Adam Hilger, Bristol.

Lindsay, D. 1984, *A Guide to Scientific Writing*, Longman Cheshire, Melbourne.
Chapter 3 includes a useful example of an *aide memoire*.

Mohan, T., McGregor, H. & Strano, Z. 1992, *Communicating! Theory and Practice*, Harcourt Brace, Sydney.
Chapter 14 is a comprehensive discussion of oral presentations and includes ways of overcoming 'speech fright' .

Patton, M.Q. 1990, *Qualitative Evaluation and Research Methods*, 2nd ed., Sage, Newbury Park, CA.

Pritchard, S. 1992, 'Friends, Romans, cost engineers ... Can we talk?' in D.F. Beer (ed.) *Writing and Speaking in the Technology Professions. A Practical Guide*, IEEE Publications, New York, pp. 212–215.

A good article which identifies the secrets and tips of five top technical speakers, thereby illustrating those factors which make the difference between a mediocre talk and a great presentation.

Richards, I. 1988, *How to Give a Successful Presentation*, Graham and Trotman, London.

Stettner, M. 1992, 'How to speak so facts come to life', in D.F. Beer (ed.) *Writing and Speaking in the Technology Professions. A Practical Guide*, IEEE Publications, New York, pp. 225-228.
Includes a helpful discussion on why a talk should cover no more than or no fewer than three main points.

Windschuttle, K. & Elliott, E. 1994, *Writing, Researching, Communicating*, 2nd ed., McGraw-Hill, Sydney.
See Part 9 for advice on preparing for public speaking, writing for the spoken word and delivery.

7 Coping With Examinations

Barass, R. 1984, *Study! A Guide to Effective Study, Revision and Examination Techniques*, Chapman and Hall, London.
Chapters 12 and 13 of this book comprehensively review steps in preparing for and completing various types of examination. Barass' book is a very helpful reference.

Burdess, N. 1991, *The Handbook of Student Skills for the Social Sciences and Humanities*, Prentice Hall, New York.
Includes two dozen useful pages on exam preparation, conduct, and review.

Clanchy, J. & Ballard, B. 1991, *Essay Writing for Students. A Practical Guide*, 2nd ed., Longman Cheshire, Melbourne.
Chapter 10, on exam essays, includes a short discussion on ways of revising for examinations.

Committee on Scientific Writing RMIT. 1993, *Manual on Scientific Writing*, TAFE Publications, Collingwood, Victoria.
Short review of examination preparation and completion strategies included as an appendix to the main volume.

Exams and Assessment (Study Skills series) 1992, Video Education Australasia.
A 15-minute video on how to prepare for an exam, dealing with stress and coping with the results.

Friedman, S. & Steinberg, S. 1989, *Writing and Thinking in the Social Sciences*, Prentice-Hall, Englewood Cliffs, New Jersey.
Chapter 13 discusses, in some detail, ten handy hints to help improve performance in written exams.

Hay, I. 1996a, 'Examinations I. Preparing for an exam', *Journal of Geography in Higher Education*, vol. 20, no. 1, pp. 133–138.

Hay, I. 1996b, 'Examinations II. Completing an exam', *Journal of Geography in Higher Education*, vol. 20, no. 3.

Hay (1996a) and (1996b) are versions of the chapter presented in this book.

Northey, M. & Knight, D.B. 1992, *Making Sense in Geography and Environmental Studies*, Oxford University Press, Toronto.
A short section on preparing for different types of examination is provided in chapter 12.

Orr, F. 1984, *How to Pass Exams*, George Allen and Unwin, North Sydney.
A comprehensive book on exam preparation and performance. Most emphasis is given to effective intellectual, physical and psychological means of preparing for examinations.

8 Referencing and Language

Australian Government Publishing Service. 1994, *Style Guide*, 5th ed., AGPS, Canberra.
A very detailed and well-laid out volume which has formed the basis for some of the material discussed in this chapter. An important reference text.

Betts, K. & Seitz, A. 1994, *Writing Essays and Research Reports in the Social Sciences*, 2nd ed., Nelson, Melbourne.
A broad review of referencing and the logic for it is contained in chapter 4.

Campbell, W.G., Ballou, S.V. & Slade, C. 1986, *Form and Style. Theses, Reports, Term Papers*, 7th ed., Houghton Mifflin, Boston.
Chapter 3 discusses the use of quotations in academic writing.

(CUTL) Centre for University Teaching and Learning and Engineering Staff. 1995, *Report Writing Style Guide for Engineering Students*, Faculty of Engineering, University of South Australia, Adelaide.

Eichler, M. 1991, *Nonsexist Research Methods. A Practical Guide*, Routledge, London.
This book provides a comprehensive review of the ways in which research practices can be sexist. The book also includes a discussion of sexism in language (chapter 7).

Harrison, N. 1985, *Writing English. A User's Manual*, Croom Helm, Sydney.
Chapter 6 'Making the text live' is an extensive review considering the uses of punctuation.

Li, X. & Crane, N.B. 1993, *Electronic Style. A Guide to Citing Electronic Information*, Meckler, Westport, Connecticut.
Electronic media such as CD-ROM, electronic journals, E-mail messages and conversations via bulletin boards present special difficulties in referencing. Although no standard has yet been established, comprehensive and helpful guidelines are outlined in this book. Drawing from the referencing principles of the American Psychological Association (APA), the book outlines forms of citation for electronic media material. Unfortunately, APA referencing style differs a little in format from that of the Australian Government Publishing Service upon which material in this chapter is based. Nevertheless, it is a simple task to make stylistic modifications to provide consistency of format.

Miller, C. & Swift, K. 1981, *The Handbook of Non-Sexist Writing for Writers, Editors and Speakers*, rev. ed., Women's Press, London.
An extensive, fascinating and useful review of sexist language, its influence, and ways to avoid it. Well worth reading.

Mills, C. 1994, 'Acknowledging sources in written assignments', *Journal of Geography in Higher Education*, vol. 18, no. 2, pp. 263-268.

Mohan, T., McGregor, H. & Strano, Z. 1992, *Communicating! Theory and Practice*, 3rd ed., Harcourt Brace, Sydney.

Peters, P. 1985, *Strategies for Student Writers. A Guide to Writing Essays, Tutorial Papers, Exam Papers and Reports*, Wiley, Brisbane.
Chapter 9 includes a detailed review of punctuation.

Index